DETERRING INTERNATIONAL TERRORISM AND ROGUE STATES

This new study challenges the widely held view that many current US adversaries cannot be deterred, maintaining that deterrence should shape US policies toward so-called rogue states and terrorist groups. The book critically assesses the "three pillars" of the Bush administration's national security policy: *missile defense*, which preoccupied the administration until 9/11; *preemption*, which became the US focus after the 9/11 attacks; and *homeland security*, which the administration embraced immediately in the aftermath of the attacks. James Lebovic argues that US policy has suffered because of severe deficiencies in US strategies. US policymakers promote defensive strategies when they should be emphasizing offensive ones, promote offensive strategies when they should be emphasizing defensive ones, and promote preemptive strategies when they should be relying upon threats to punish countries for acquiring or using illicit weaponry.

Deterring International Terrorism and Rogue States ultimately establishes that inadequate offensive and defensive strategies have led US policymakers to pursue open-ended policies without adequate concern for resource trade-offs, overreach, and unintended consequences.

This book will be of great interest to students of US foreign policy, national and international security, terrorism, and international relations in general.

James H. Lebovic is on the faculty of the Department of Political Science and the Elliott School of International Affairs at the George Washington University. He has published widely on international conflict, deterrence, military spending, and defense strategy and policy. He is the author of two prior books: *Deadly Dilemmas: Deterrence in US Nuclear Strategy* and *Foregone Conclusions: US Weapons Acquisition in the Post-Cold War Transition*.

CONTEMPORARY SECURITY STUDIES

DETERRING INTERNATIONAL TERRORISM AND ROGUE STATES

US national security policy after 9/11

James H. Lebovic

Routledge
Taylor & Francis Group

LONDON AND NEW YORK

First published 2007
by Routledge
2 Park Square, Milton Park, Abingdon, Oxon OX14 4RN

Simultaneously published in the USA and Canada
by Routledge
270 Madison Ave, New York, NY 10016

Routledge is an imprint of the Taylor & Francis Group, an informa business

Typeset in Times New Roman by
Florence Production Ltd, Stoodleigh, Devon
Printed and bound in Great Britain by
Antony Rowe Ltd, Chippenham, Wiltshire

British Library Cataloguing in Publication Data
A catalogue record for this book is available from the British Library

Library of Congress Cataloging in Publication Data
Lebovic, James H.
Deterring international terrorism and rogue states: US national
security policy after 9/11 / James H. Lebovic.
p. cm.—(Contemporary security studies)
Includes bibliographical references and index.
1. United Nations—Military policy.   2. National security—
International cooperation.   3. Deterrence (Strategy).
4. Terrorism—Prevention—International cooperation.
5. World politics—21st century.   I. Title.   II. Series.
UA10.5.L428 2006
355′.0335—dc22   2006016560

ISBN10: 0–415–77143–9 (hbk)
ISBN10: 0–415–77144–7 (pbk)
ISBN10: 0–203–08900–6 (ebk)

ISBN13: 978–0–415–77143–6 (hbk)
ISBN13: 978–0–415–77144–3 (pbk)
ISBN13: 978–0–203–08900–2 (ebk)

TO THE MEMORY OF MY FATHER

CONTENTS

ix

CONTENTS

FIGURES

ACKNOWLEDGMENTS

This book took form over a five-year period. It started as an article on deterrence and missile defense that I drafted just before September 11, 2001 and then revised and published in the months that followed. My effort to assess US national security policy in light of deterrence principles would likely have ended there but for shifts in US strategies to combat terrorism and proliferation that convinced me to extend my analysis.

As this work evolved, it benefitted from the suggestions and criticisms of a number of people within the George Washington University community—Debbi Avant, Steve Balla, Mark Esper, Henry Farrell, Jim Goldgeier, Eric Lawrence, Henry Nau, Marc Raskin, Jeff Richelson, Sam Rothenberg, David Shambaugh, Lee Sigelman, Erik Voeten, and Langche Zeng. Two people deserve special recognition—Michael Gee, for his skilled research assistance, and my colleague and friend, Bob Stoker, for saying, at a receptive moment, "you really should write a book about this." I note that those I have thanked are not equally comfortable with the contents of this book; I owe a special debt of gratitude to readers who were willing to make this a better book, because it was *my* book. I also wish to thank Robert Jervis and R. Ned Lebow for their prolific output of works that have profoundly affected my own thinking on the subject.

I am grateful, as well, to Andrew Humphrys at Routledge for recognizing a potential audience for my work and the anonymous reviewers at the press who highlighted deficiencies in my manuscript and made helpful recommendations. Deserving recognition, too, are Bruce Russett and reviewers at the *Journal of Conflict Resolution* for their useful comments on a draft of my missile defense article, which has been reworked for inclusion as a chapter in this book.

Finally, I owe the greatest debt to Holly—related by marriage to a book project, and all that comes with it. This includes an intrusive security search at a foreign airport because I chose to take my work on vacation with us. I would apologize but my hunch is that would not be enough.

ABBREVIATIONS

ABM	anti-ballistic missile
AD	assured destruction
CIA	Central Intelligence Agency
FARC	Revolutionary Armed Forces of Colombia
GDP	gross domestic product
HEU	highly enriched uranium
IAEA	International Atomic Energy Agency
LTTE	Liberation Tigers of Tamil Eelam
MIRV	multiple independently targetable reentry vehicle
NATO	North Atlantic Treaty Organization
NMD	national missile defense
NPT	Nuclear Non-Proliferation Treaty
PKK	Kurdistan Workers' Party
PLO	Palestine Liberation Organization
SAFP	single-attack failure probability
START	Strategic Arms Reductions Treaty
SWAT	special weapons and tactics
WMD	weapons of mass destruction

1

INTRODUCTION
Deterrence in a changing world

Most Americans vividly recall where they were and what they were doing on September 11, 2001 when they watched the twin towers of the World Trade Center in New York collapse in flame in the deadliest foreign attack ever on US soil. This personalized tragedy thrust Americans reluctantly into the unfamiliar world of international politics. It is understandable, then, that they are prone to overstate how much the events of that fateful day imposed new demands upon US foreign policy. Now, for instance, US adversaries are commonly presumed to pursue the same nefarious goal—inflicting maximum damage upon the US and to its interests, at any and all cost. Ungrounded assumptions are not confined to the public. The George W. Bush administration added to the confusion by repeatedly labeling Iraq (under Saddam Hussein), Iran, and North Korea "rogue states," treating them as co-plotters in a global conspiracy (e.g. the "axis of evil"), and claiming proof of cooperation between Iraq's former leadership and the al-Qaeda terrorist network to suggest that Saddam Hussein was behind the World Trade Center attack.[1] The events of September only strengthened the administration's view that the US is now facing adversaries that cannot be deterred—that is, dissuaded by potential costs—from pursuing their goals.

It is tempting to assume that deterrence principles *are* obsolete given their origins in the nuclear realities of the Cold War era. The superpowers were famously compared then to "scorpions in a bottle"—neither was secure as long as the other lived, nor could kill the other without first being stung fatally in retaliation. This oppressive reality meant that the Cold War was a "golden age" for abstract thinking about deterrence and exploring its policy implications: the belief that the Soviet Union could be "deterred"—in some sense of the term—with the right US strategies was held almost universally within the US policy community. But a consequence of this perceived association between deterrence and the US-Soviet nuclear stand-off was that deterrence principles seemed increasingly antiquated with the waning of Cold War tension.

That challenges of the present and dissimilarities with the past have caused policymakers to doubt the utility of deterrence principles is unfortunate. Whereas these principles permitted *some* restraint in dealings with adversaries, restraint is no virtue in the current view. Relying upon the logic that "all that changed with September 11," US policymakers have pursued policies that require the US to do everything without concern for resource trade-offs, overreach, and the unintended consequences of policy. The "war on terrorism" has become the rallying cry for the US as it races off in multiple directions in the pursuit of open-ended policies.

What is needed, I argue, is the sense of balance and proportion that derives from attending to first principles and drawing from policy assumptions that governed US thinking in the Cold War period. I maintain that deterrence theory is not a Cold War relic, and that it should shape policies toward even those US adversaries—the "hard cases" of rogue states and terrorists—that supposedly cannot be deterred. I assert that deterrence is far more robust than policymakers concede, and that policy guided by the false premise that deterrence does not work will justify extreme and imprudent measures in a "one-size-fits-all" foreign policy that will only increase the chances that US efforts will fail.

The contemporary policy debate

Critics are on the defensive in the US national security debate and have generally deferred to administration judgments. The "debate" over US policies toward so-called rogue states has focused, in the main, upon the execution of policy rather than upon its assumptions. The post-war debate over the 2003 US invasion of Iraq reduced first to whether US intelligence on Iraq's weapons of mass destruction (WMD) was good or bad, and then who was responsible for the flawed data that were used to justify US intervention. The issue of whether such weapons in the hands of Iraq necessitated US military action had been settled; and the option of preemptive action against Iran and, perhaps, North Korea remains on the table. Missing from the contemporary policy debate is due appreciation that reckless US preemptive policies are a greater threat to the US than the mere possession of nonconventional weapons by rogue states. These policies might not work and can actually undermine deterrence, inviting the consequences that preemption advocates appear to fear. Unexamined assumptions also pervade US defensive plans to counter rogue states. Policymakers fail to acknowledge that ambitious US efforts to defend against rogue-state missile attacks can exacerbate deterrence problems if pushing US adversaries to adopt destabilizing countermeasures or to exploit the vulnerabilities of the defense and that, regardless, the effectiveness of US defenses will depend upon a US threat to punish offenders through a deterrence strategy.

The assumptions underlying US counter-terror strategies similarly escape scrutiny. The current administration has opted for the offensive strategy of "taking the war to terrorists" by attacking their leadership, financial centers, networks, and support base. The guiding assumption for US policymakers is that "playing not to lose" guarantees defeat. The value of offensive tactics appears self-evident: why permit the adversary to retain the initiative, that is, to decide when, where, and what to attack? Why would the US forego opportunities to build cooperation with intelligence and police units abroad, to capture and kill individuals who provoke or plan terror operations before an attack occurs, to fight enemies abroad rather than within US borders, and to train and work with foreign militaries so US forces are not required to be everywhere and do everything? For no good reasons—at least, so it seems. But US policies could benefit enormously from a shift from offensive to defensive priorities. Despite US policy preferences, offensive tactics have their limits, can be self-defeating, and can create hardships for a fragile US defense. Homeland security need not be subordinate to an offensive strategy: offenses can be replaced by defenses that are designed, in part, to *deter*.

In exploring the contemporary application of deterrence, I focus upon the "three pillars" of the Bush administration's national security policy: *missile defense*, which preoccupied the administration until September 11, 2001; *preemption*, which became the US focus with the September 11 attacks; and *homeland security*, which the administration portrayed as more a natural response to threat than an aspect of policy that must be reconciled with the other pillars. I argue that, given the neglect of deterrence, policymakers: (a) promote a defensive strategy when they should be emphasizing an offensive one; (b) promote an offensive strategy when they should be emphasizing a defensive one; and (c) promote an offensive-denial strategy (aimed at physically disarming an adversary) when they should be promoting an offensive-punishment one (based on coercing an adversary). Viewed in terms of the pillars of policy, then, I argue that policymakers: (a) exaggerate the value of offenses relative to defenses when making the case for a preemptive strategy against terrorists; (b) exaggerate the value of offensive-denial relative to offensive-punishment when supporting a preemptive strategy against rogue states; (c) exaggerate the value of defenses relative to offenses (i.e. the threat of retaliation) when arguing for a national missile defense (NMD) system; and (d) depreciate the value of homeland defenses by lavishing attention on offensive alternatives.

At least in theory, offenses and defenses are complementary. It is a truism in competition that a good offense makes for a good defense, and the reverse. But offenses and defenses are not always compatible, and each can undermine the other. Bad offenses—those that are poorly suited for the intended task—can compound problems for the defense by increasing its challenges and absorbing scarce resources. In turn, bad defenses can

3

provoke adversary reactions that create and aggravate problems for the offense. Simply put, bad offenses and defenses are endemic to the current US policy approach. US offenses subvert a deterrence strategy by sapping resources that could be used in homeland security, by increasing the fears of rogue-state leaders, which could cause them to take destabilizing countermeasures, and by motivating and (perhaps even) enabling terrorists to attack. US defenses are inadequate for the problem at hand and, in the case of missile defenses, can weaken deterrence if viewed by adversaries as a threat.

Deterrence: Cold War-era limitations

Deterrence results when a party forgoes action because its costs outweigh the benefits. This conception, which associates deterrence with an ability to punish another for its transgressions, guided US national strategy in the Cold War years. As a result, the US sought to develop the capabilities to impose costs upon the Soviet Union, should it choose to act against the US and its interests in various parts of the world. As importantly, the US sought to convince the Soviet Union that the US meant what it said and would stand by its commitment to defend those interests should the need arise. This task seems straightforward when expressed in these terms. But US policymakers appreciated that the Soviets might doubt US resolve given the horrendous consequences of nuclear war, which the US *did* want to avoid, and US commitments to many countries that were conceivably of marginal importance to US security. US strategists rose to the challenge. Academics and policy intellectuals tweaked, augmented—and even transformed—deterrence principles to address critical *theoretical* challenges within a Cold War context.

This does not mean that Cold War-era deterrence principles were unproblematic: to the contrary, they were highly problematic. They were sometimes based on heroic assumptions about the adversary—its ability to think dispassionately, process information, and make the "right" decision under the most challenging of conditions—literally, seconds in which a national leader would decide the future of the planet. We should not forget the near-misses of the Cold War period at critical junctures, misunderstandings and organizational behaviors that could have led to war, the debt we owe to the wisdom and caution of particular US and Soviet leaders in moments of crisis, and the extent to which nuclear deterrence was itself a learning process. The Soviets did not take to principles of deterrence that contradicted long-standing Soviet bureaucratic traditions and historical lessons any more easily than many US policymakers did. For that matter, US and Soviet relations through much of the Cold War did not conform entirely, and even primarily, to deterrence principles—a fact that was sometimes lost on the participants.

An understanding of the current security challenge must reflect an aware-ness, then, that deterrence is and has always been a problematic strategy. The many problems of applying deterrence principles to Cold War-era behavior remain relevant today.

First, these principles can overstate the malevolence and ambitiousness of contesting parties. The superpowers were depicted, by all sides in the deterrence debate, as opportunistic—constantly probing for weaknesses and conditions that could be exploited to foster the other's demise. But it was never clear what the parties could gain from neutralizing the other, much less whether this was indeed the US or Soviet goal. The historical record actually shows that both superpowers fell well short of the level of mutual aggression that is implied by the deterrence model. The US chose not to attack the Soviet Union even in the early 1950s, a period of over-whelming US nuclear supremacy; and the Soviet Union chose not to jump through the "window of vulnerability" that hawkish policymakers of the 1980s believed had been opened by the Soviet deployment of a large number of land-based missiles with multiple warheads aimed at fewer US land-based missiles. For that matter, neither party tested the other directly by initiating conventional military confrontations. As Lebow and Stein (1994: 357) conclude in their impressive study of US–Soviet crisis decision-making:

> it was not only the absence of *opportunity* that kept the peace, but also the absence of a strong *motive* for war. Without a compelling motive, leaders were unwilling to assume the burden and responsi-bility for war, even if they thought its outcome would be favorable.

Put plainly, "the reality of deterrence helped to restrain leaders on both sides, but their relative satisfaction with the status quo was an important cause of the long peace."

Second, policymakers can mistakenly assume that coercion is an effective source of influence. The escalations of conflict that some deterrence theorists presume can usefully signal an actor's stakes and capabilities in a crisis can reinforce the nonlogical tendencies of the target, and provoke it. Whereas deterrence theorists suppose that threats are a stabilizing influence, threats can have the opposite effect. The psychological aspects of this process have been subject to considerable research. It has established, most notably, that deterrence threats can create a self-fulfilling prophesy when these "defen-sive" actions are mistaken for aggressive gestures. A party that fears another, is unlikely to view it more charitably after being threatened by it even if meant to convey that compromise is preferable to conflict. The pressure toward conflict is that much greater when policymakers yield to a variety of other (mutually reinforcing) psychological tendencies. These include fore-closing options prematurely when considering policy alternatives, ignoring policy trade-offs and assuming mistakenly that preferred policies have only beneficial effects, misreading and misapplying information when evaluating alternatives, overconfidence and wishful thinking, bolstering of preferred policies by gathering supportive evidence and ignoring disconfirming evidence, and underestimating constraints that limit an adversary's latitude for choice (see, e.g. Jervis 1976, 1982/3). The likelihood that deterrent threats will backfire only increases when their targets reject the status quo and view it as politically, economically, or militarily untenable: "if leaders are driven less by the prospect of gain than they are by the fear of loss, deterrent policies can provoke the very behavior they are designed to forestall by intensifying the pressures on the challenger to act" (Lebow 1989: 27). The desire to fight "fire with fire" is that much greater when leaders see the world, and act toward it, through clumsy government bureaucracies and face domestic challengers who will capitalize on signs that those leaders are soft on issues of national honor or security. The propulsion toward conflict can worsen in crisis (see Jervis *et al.* 1985).

Third, deterrence theory focuses attention on short-term actions and responses when short-term successes can produce "failures" with time. Just as US backing of Islamic militancy to counter the Soviet presence in Afghanistan seems ridiculously short-sighted in light of contemporary events, the short-term US success at forcing Soviet missiles out of Cuba in 1962 can be viewed as a long-term failure for US policy insofar as the missile crisis helped propel the intense Soviet nuclear buildup in the years that followed. The issue of Soviet missiles in Cuba was moot once the Soviets could deliver missiles over intercontinental distances and fire missiles from submarines located off US shores.[2]

Fourth, policymakers assume, often wrongly, that they share their counterparts' understanding of the "situation." If the sides hold incommensurate views of the conditions and stakes in a conflict, coercive actions might serve only to reinforce dangerous misconceptions. Ultimately, leaders will not be deterred if all possible futures look bleak—if they believe there is little left to lose by testing the will of an adversary and by responding to threats with force. In attacking Kuwait in 1990, it appears that Saddam Hussein actually saw *himself* as the injured party—that he believed he was the "defender," not the "challenger." This is important because the "defender" might feel that it has little choice but to act, even a moral obligation to do so.

Fifth, policymakers assume that "deterrence problems" are actually deterrence problems. Foremost among the pretenders are "compellence" problems, where compellence is understood to involve attempts by a party to *change* the status quo through coercion. The danger is that policymakers will exaggerate the usefulness of threats against adversaries that are loath to convey weakness and renounce policy options. (This issue receives additional attention in Chapter 2.)

Sixth, policymakers can misread deterrence principles to require a linkage between issues and problems in diverse parts of the world. Concerns that the Soviet Union or US allies might doubt the resolve of the US government—especially, its willingness to do the unthinkable and stand by its suicidal threat to retaliate for a Soviet attack—provoked an obsessive US desire not to appear weak or irresolute in crises around the globe. Cold War-era policymakers disagreed over exactly how issues were linked but they agreed that issues *were* linked—that all US behavior would be read by adversaries and allies alike for underlying messages about US intentions and capabilities. Case study analysis suggests that the messages of US victories and defeats throughout the Third World were not read by the Soviet Union as the US had predicted (see Hopf 1994). Still, linkage effects continue to preoccupy the US as it pursues an ambitious strategy with the assumption that the US confronts a monolithic, global nemesis in the "axis of evil" and terror threat.

Finally, deterrence principles were too easily manipulated by US policymakers to serve their preferences and, specifically, a pre-occupation with weapons capabilities and deployment modes. With the assumption that quality and versatility were paramount, the US pushed ardently to increase the accuracy of US offensive forces and to diversify the means for delivering US nuclear weapons against their targets. Consequently, the assumptions behind US policies were far more inconsistent and undeveloped than would be guessed from the sophisticated tactics and technologies that were linked to abstract deterrence strategies (Lebovic 1990). Policymakers adopted policies to promote deterrence that could have undermined it.

In the end, deterrence principles were perhaps *too compelling*, for they could justify unrestrained actions with the false assumption that conflicts will be resolved to the benefit of the party that talks loudest and waves the bigger stick. A deterrence strategy is not a safety net for incautious and injudicious behavior; it does not reduce to determining what an adversary values and threatening it forcefully. A lesson of the Cold War is, instead, that deterrence can work when threats are implicit, what is unacceptable conduct comes to be understood, adversaries appreciate the limits to each other's objectives, and opportunities for compromise are maintained and created (Lebow and Stein 1994). In other words, deterrence can work when policymakers allow situations to speak for themselves and leave the door open for negotiated resolutions of disputes.[3]

These Cold War lessons remain instructive because much about the world did not change with the end of the Soviet threat. Indeed, the principles that governed US–Soviet relations perhaps better fit today than the Cold War period. The US is far better positioned now than in the Cold War years to inflict prohibitive costs upon an offending state, through a variety of conventional and nonconventional means, and to limit the resulting costs and risks to the US and its allies. The reduced costs to the US of intervening worldwide have only strengthened deterrence by increasing the credibility of US threats to act when provoked—whether against states that target the US or its allies with WMD or support terrorist actions against those same targets. True, US adversaries might doubt that the US will act on its threats. But deterrence does not require that adversary leaders believe assuredly that the US *will* respond—only that it *might* respond—to an attack (Jervis 2003: 321–3). The irony, then, is that US interest in deterrence (and deterrence theory) declined as many strategic challenges that troubled the critics of mainstream US policy lessened considerably into the post-Cold War period.

Similar lessons apply in crafting a deterrence strategy against terror attacks. A defensive (denial-based) deterrence strategy, which plays to the adversary's existing perceptions of benefits and costs, could be more productive than a strategy that seeks to destroy the terror infrastructure abroad or signal that terrorists will be punished for their transgressions. It is easier to convince terror adversaries that their plans are too ambitious or risky, given US defensive capabilities, than to destroy these groups or convince them to renounce their goals because the costs of being a terrorist outweigh the benefits.

The potential applications of deterrence are not lost entirely on US policymakers. The authoritative, four-year US defense planning document (the *Quadrennial Defense Review*), released in early 2006, uses the word "deter" (or its variants) over fifty times in calling broadly for "tailored deterrence for rogue powers, terrorist networks and near-term competitors" (US DOD 2006a: vi). Yet deterrence remains an ancillary theme throughout the document, which conveys US options against rogue states and terror groups with terms such as "prevent," "defeat," "defend," "destroy," and "disable." Deterrence, to the extent that it matters in US relations with rogue states, ensues therein from an ability to employ overwhelming force to "deny" these states any *benefits* from an attack (US DOD 2006a: 27)—not from the *costs* that US forces can impose upon these states should they attack. That deterrence in any form applies directly to terrorists is not considered. Although deterring terrorists is mentioned in the *National Military Strategic Plan for the War on Terrorism* (US DOD 2006b), which offers a blueprint for a global US war on "violent extremism," the document centers almost entirely on a military strategy for denying terrorists the means to carry out attacks.

An appreciation of the value of deterrence might grow with time; and there is evidence that the Bush administration, with its lingering effort in Iraq, became less sanguine about the utility of force in dealing with the new threats. For that matter, the evolving US response to events and their rapid pace could date this analysis with (or before) a change in administration. As I write, in the fall of 2006, the Iranian nuclear program is before the UN Security Council for possible action; the North Korean nuclear program is progressing, perhaps rapidly, in secrecy; Hamas, after engineering a surprise electoral victory in the Palestinian legislative elections, is struggling in governance; Israel has withdrawn from Lebanon after intense combat with Hezbollah forces; the fighting in Iraq is apparently escalating toward civil war; terror attacks on government, civilian, and foreign military targets are becoming more frequent in Afghanistan; and the al-Qaeda terror organization is still in business, as are its most notorious leaders. But an increase in the pace of events and dramatic new developments would only strengthen the rationale for this book because it is written in part as a cautionary tale. The book should be read as an admonition to policymakers who single-mindedly pursue simple and/or ambitious solutions to complex policy problems. Of course, the book should be read as well for its insights on contemporary policy. In all likelihood, the book's main arguments will not soon be overtaken by events and the cornerstones of Bush administration policy will inform if not structure US policies into the future. Indeed, in early 2006, when many commentators claimed that the administration was reevaluating its commitment to preemptive principles, the administration reaffirmed those principles with the release of the President's *National Security Strategy* statement (Bush 2006), and backed its words with actions, in the form of contingency plans for possible attacks on the Iranian nuclear infrastructure (Arkin 2006).

The book

For the reader more interested in some facets of the deterrence problem than others, the four "issue" chapters (Chapters 3–6)—focusing, respectively, on offensive strategies against rogue states, missile defense, offensive strategies against terror groups, and homeland security—are written to be read out of order. Whatever the order in which the reader proceeds, my hope is that they will appreciate the value of addressing these issues within a deterrence framework. It can highlight trade-offs that are required in pursuing foreign policy objectives and can disclose similarities among a variety of issues to permit a deeper understanding of policy problems.

The book is designed as follows. Chapter 2 revisits the central assumptions of Cold War-era deterrence theory and suggests, with a light touch, how these assumptions apply to the present. Chapter 3 examines the usefulness of punishment and denial strategies when taking the offensive against rogue states. Reflecting on lessons learned from the behavior of Saddam Hussein's Iraq, Iran, and North Korea, it explores the assumption that rogue states are unresponsive to punishment. It then assesses the benefits and costs of denial strategies based in preemptive (and preventative) war and less intrusive measures such as embargos and blockades. It concludes that punishment is far more useful than critics concede and that preemption is far more problematic: reckless US preemptive policies against rogue states could be a greater threat to the US than the reckless use by rogue states of WMD. It concludes further that denial options might work best, not as a substitute for deterrence, but when serving an economic punishment strategy or a military strategy that threatens disarming retaliation in the event of a rogue-state attack. Chapter 4 assesses the case for a US NMD given a variety of scenarios in which missiles could be launched against the US or be used to threaten the US for coercive benefits. The chapter surmises that the value of a missile defense of any type and scale is low because it is unlikely to enhance global security above levels offered by retaliatory deterrence in the threat to punish adversaries for an attack. Indeed, it concludes that missile defenses can exacerbate deterrence problems. Chapter 5 explores the utility of offensive (punishment and denial) strategies, applied to the global terror threat, by assessing three interrelated perspectives that policymakers and scholars embrace to understand the terror problem. It maintains that offensive strategies can succeed, whether based on coercion or efforts to destroy or dismantle the terror infrastructure. It also recognizes, however, that these strategies have major liabilities and could well be unsuccessful when applied from a distance. Chapter 6 returns to the defense in examining the deterrence logic behind efforts to promote homeland security against terrorist attacks. It argues that the attacker's options are constrained and, therefore, that deterrence opportunities are available to governments that reject the notion of the risk-taking terrorist—out to destroy at all cost—in favor of the Cold War assumption of a rational adversary that chooses carefully when, what, and where to strike. It concludes, then, that homeland defense can serve a (denial-based) deterrence strategy and is unfairly discounted when viewed as the "last-line" of defense—as the final recourse when all else (an offensive strategy) has failed. Chapter 7 summarizes the advantages of applying deterrence principles to the contemporary world. It also argues for maintaining perspective when assessing the actual magnitude of current US national security challenges.

2

VIEWING THE PRESENT THROUGH THE PAST

That policymakers and analysts have neglected the old strategic principles and debates is somewhat understandable, for much of the reasoning was hypothetical, arcane, and counterintuitive. It is less understandable in light of the continued relevance of issues that were central to Cold War-era deterrence theory. Among these issues are the emphasis to be accorded punishment relative to denial, and offense relative to defense, in US strategy, the benefits and liabilities of assuming that US adversaries are rational, the challenges to deterrence should the US try to "extend deterrence" to encompass third parties, and the limits of deterrence if it is understood to require that the US compel adversaries to concede ground.

Constitutive concepts

During the Cold War, US thinking about deterrence was based on pairings of concepts across two sets: (a) punishment and denial, and (b) offense and defense. Whether the pairing of offense and punishment, offense and denial, or defense and denial prevailed in policy was not always clear and even a dominant pairing in government doctrine did not bring coherence to US national security policy. But these pairings were key to the evolution of US nuclear strategy, as is apparent when examining their constitutive concepts.

Punishment and denial

Central to US deterrence reasoning was the assumption that the Soviet Union would not attack the US if its retaliation were quick, certain, and devastating. This logic was the basis, in the 1960s, of US "assured destruction" (AD) doctrine. AD doctrine had the US preparing to withstand a worst-case, Soviet first strike—a surprise, all-out attack—and then render damage in return (a second strike) that met established AD criteria. By these criteria, between a quarter and a third of the Soviet population would have been killed and two-thirds to three-quarters of Soviet industries would have been destroyed in a US retaliatory attack. The high toll reflected the US supposition that the Soviet Union could be dissuaded from attacking the US with nuclear weapons if the costs of an attack were sufficiently prohibitive. With offsetting increases in Soviet nuclear capability soon to follow, the assumption was given bilateral form, in the principles of Mutual Assured Destruction, the mainstream perspective of the Cold War period. It supposed that a stable deterrence relationship between the US and Soviet Union was possible as long as each superpower retained the capability to render "punishment" for an attack—of any type and scale—by destroying the attacker. The emphasis then was on using nuclear weapons in retaliatory "counter-value" strikes that targeted enemy cities and industries.

Deterrence theory was not a cohesive set of arguments, though. It was a collection of often competing perspectives that developed through intense debates over the risks and damage that US adversaries should be made to incur and the strategies with which adversaries would pursue their goals. "Doves" embraced AD logic and the accompanying faith that the Soviet Union would do the right thing in the face of temptation. But "hawks" articulated a "war-fighting" position. They supposed that the Soviet Union—given its grand ideological imperatives and implacable hostility toward the US—would exploit existing military or political opportunities for gain. Following Kahn (1961), war-fighters were concerned that a "limited" Soviet attack on the US would leave the US with no rational options but to surrender or, instead, to commit suicide by retaliating against the Soviet Union and inviting Soviet retaliation in return. War-fighters (along with some pessimistic AD proponents) were driven, then, to search for options that would deter the Soviets, by reducing their incentive to launch such an attack, and/or forestall national suicide should war occur. Different strategists gravitated toward different solutions: manipulating risk with limited nuclear strikes, fighting a protracted nuclear war by focusing attacks on military targets, or some combination of the two. Of these, the "pure" war-fighting strategy posed the greatest challenge to AD doctrine. In pure war-fighting, the purpose of a US second strike (and/or US defenses) was to "deny" the enemy gains from attack, not to punish it for an attack: deterrence could be achieved by the side that could neutralize its opponent's ability to acquire advantages through war or that could *prevail* at the higher level of conflict (so that the opponent could not achieve gains through escalation; see Gray and Payne 1980). Here, the emphasis was on using US weapons to disarm the Soviet Union through "counter-force" attacks (see Freedman 2004: 36–40; Jervis 1984; Snyder 1961).

To (AD) defenders of the status quo, war-fighters carried their argu-ments to illogical extremes with policy recommendations that rested on false premises. Among their failings, war-fighters were said to exaggerate the meaningfulness of apparent US and/or Soviet capability advantages, disparities between US and Soviet intentions, the ability of the superpowers to act in their best interests in the "fog of war," and the US ability to control the consequences of a nuclear war once it had erupted. Nevertheless, US policymakers of the 1970s and 1980s increasingly deferred to war-fighting principles and developed highly elaborate doctrines and retaliatory options to fight protracted nuclear wars. The policy shift was assisted by computational advances, formal decisional tools, improvements in weapons accuracy, efficiency, and responsiveness, a vast US research and development base, and the absence of direct superpower conflict that could reveal deficiencies in US doctrine.

Although the nuclear debate largely centered, then, on whether the Soviet Union was deterred by threats of societal *punishment* (the imposition of costs) or instead by the *denial* of gains from an attack (the absence of benefits), clear winners and losers seldom emerged from the policy process. Deployment and employment policies were rarely true to one set of principles; in fact, policies were counterproductive given these principles.

The disconnect between policies and their theoretical justifications was illustrated by US strategic missile deployments. Land-based (intercontinental ballistic) missiles intended to fight a nuclear war should presumably be placed at distances from US cities to limit collateral damage from a Soviet first strike. The US had nothing to gain and everything to lose from subjecting its citizens and industries to attack. Conversely, US missiles for executing AD doctrine should arguably have been placed in the middle of cities to show the Soviet Union that the US meant business. What better way to convince the Soviet Union that the US would retaliate for an attack than placing itself in a position where it had nothing further to lose by retaliating against Soviet cities? After all, did it matter if the US population was destroyed in the first rather than the third-wave of attack, when the Soviet Union retaliated for US retaliation? At most, deploying missiles in rural rather than urban areas amounted to buying but an additional hour of life for people residing in those cities given the speed with which Soviet and US missiles could reach their targets. Put simply, if the US wanted to communicate that it intended to deliver on its "murder-suicide" pact with the Soviet Union, then making US cities *more* vulnerable to any Soviet first strike—not less vulnerable—was the solution. Despite its AD logic, the US chose to lessen its exposure to an attack.

The problem then is that broad thinking at the level of doctrine, whether war-fighting or AD, never translated neatly into US war plans. This was even true of concepts that were central to these doctrines. It makes sense in the abstract to distinguish between urban (punishment) and military (denial) targets for pursuing counter-value and counter-force strategies but these targets are hard to separate in practice. It seems a simple matter to determine whether a government building or industry is a military or civilian target depending on the kinds of activities that are conducted there.

But this assumes that a government building or industry has just a single function and that agencies and industries that perform different functions are physically separated from one another.

That US policymakers once found it hard to stick to their principles does not diminish their importance, in the present, for developing sound strategies. US policymakers must weigh the merits of punishment and denial strategies when formulating policies to combat current US adversaries.

Policymakers have promoted a (denial-based) preemption strategy that assumes that rogue states must be forcefully disarmed because they *cannot be deterred*. But a preemptive strategy is challenging to implement given the difficulties and (financial and human) costs of locating and destroying targets. Some states are unlikely to be disarmed fully or permanently by a US offensive (or deterred, then, from acquiring illicit weapons by a fear that these weapons will be destroyed before they can be used). Thus, preemption will not end the threat from these weapons; and coercion might still be required to obtain post-attack access to the facilities and records of the targeted country to determine the residual state of its weapons program. That US policymakers must live in a world in which the principles of punishment trump the principles of denial is not necessarily an unappealing prospect. A punishment strategy can rely upon a combination of political, economic, and/or military (denial) instruments—including blockades and embargos—to increase the negative (political, economic, and military) consequences for states seeking nonconventional capabilities.

The limits of denial relative to punishment strategies also haunt US efforts to build a US NMD system to thwart a rogue-state missile attack. Although missile defenses are promoted by US policymakers partly as a denial-based alternative to deterrence, the costs and technological challenges of even a working NMD system could place the US at a strategic disadvantage and will not end US reliance on deterrence threats to punish adversaries for their transgressions. Indeed, in a fundamental sense, inflicting imposing punishment on adversaries is the operative intent of the system. Whereas an effective missile defense can deny *gains* to an attacker by preventing it from destroying a US city, it can also impose opportunity *costs* on an attacker in the form of a lost missile investment. In other words, a missile prevented from reaching its target is unavailable for the attacker's other purposes. These include deterring a conventional or nuclear attack, attacks on the same target under more favorable circumstances, and strikes on alternative targets. Put simply, an Iranian missile that is fired unsuccessfully at a US target is a missile lost in a war with Israel. To say that denial is the sole basis of deterring a missile strike, then, is to presume a single-minded adversary, a foe that has *but one goal and one moment to accomplish it*. If an adversary has two objectives—say, to destroy a US city and to stay in power—the distinction between denial and punishment collapses. If accomplishing the first objective brings massive US retaliation that weakens the regime, the achieving of the second objective is in doubt. Consequently, the regime is effectively punished if unable to accomplish that goal.

Therefore, when skeptics claim that a reprisal (punishment) strategy does not deter, what they really mean is that reprisal does not work because: (a) adversary leaders, living in the moment, care only about inflicting as much damage as possible on the US; or (b) the adversary does not care about the *specific* costs that the US can or will impose. If the latter is true, conditions call only for a refined retaliatory (punishment) strategy. Indeed, that very conclusion was drawn by many AD critics who tried hard to understand what it would take *to deter* a supposedly ruthless Soviet leadership. For them, the worst-case assumption was not that Soviet leaders were undeterred by retaliatory threats but rather that Soviet leaders were undeterred by the kinds of threats that deter US leaders. It was this alleged asymmetry in US–Soviet values that led US strategists in the 1970s and 1980s to insist that US nuclear targeting imperil, among other things, the Soviet leadership, putting its existence and political interests at risk.

Punishment is even effective against some terrorist groups. Although it is generally assumed that terrorists are resistant to punishment because they do not have "values" (i.e. human or material assets) that can be placed at risk or at least targeted easily for destruction, terrorist groups are susceptible to punishment when a basis of compromise exists between these groups and opposing governments. If terrorist groups choose, for example, to limit their violence so as not to bring a third-party (e.g. the US) into a conflict or to invite wholesale retaliation by a government (e.g. Israel), then these groups concede, by their behavior, to a deterrence logic. Punishment is also applicable to terror groups that value their capability to conduct an operation. When terrorists are uninterested in protecting territory and citizens from attack but are driven by ideology, religion, or some overriding sense of purpose to elevate a "cause" above all else, they must also safeguard assets—leaders, bases, finances, plans, and weapons—that are used in pursuit of the cause. When terrorists waste or lose their capabilities, they are also punished.

The upshot is that even if punishment does not directly affect terrorist calculations, it can work indirectly through the principles of a denial strategy. Simply put, terrorists are punished when they are denied their goals because: (a) they have lost capabilities with the destruction of their weapons, finances, bases, or leaders; or (b) they imprudently expend their capabilities through poor planning or wasteful expenditures of resources on unimportant targets. In this important sense, efforts by terrorists to protect or husband their resources are effectively concessions to coercion. Terrorists are *punished*, then, when they are *denied* their objectives through US offensive strategies that target these assets (e.g. terrorist leaders) and defensive strategies that make it more difficult for terrorists to carry out

15

their attacks without prematurely exposing or inefficiently using those assets. If terrorists are dissuaded from acting—out of a fear of being attacked or of premature detection of an attack plan—the case is strengthened for (offensive or defensive) *denial-based* deterrence strategies.

Certainly, governments work at a disadvantage when pursuing offensive, denial-based strategies against terrorists. When on the offensive, governments must contend with difficulties in locating elusive terrorist leaders and their assets, distinguishing terrorist from civilian targets, increasing terrorist support through indiscriminate attacks on the terror infrastructure, and disabling terror groups when new leaders can replace assassinated ones or the structure of the terrorist organization can change to accommodate its hostile environment. These problems argue for shifting the US policy emphasis toward homeland security. Although this defensive, denial-based approach benefits terrorists by allowing them to choose when, where, and how to attack, the advantage can shift toward the government when it is best positioned to defend what terrorists seek to attack or when the government can manipulate the risks and costs to terrorists of an attack. If terrorists accommodate the defense by halting an attack or altering it in ways that reduce its destructiveness, governments reap rewards from a (denial-based) deterrence strategy.

As detailed in the chapters to follow, policymakers must appreciate the usefulness and limitations of punishment and denial-based deterrence strategies when applied to contemporary security problems. They must also acknowledge that denial instruments (e.g. embargos) can serve a punishment strategy.

Offense and defense

Whether US strategy should rely on punishment or denial depends on the effectiveness of US offenses and defenses, in absolute and relative terms. The issue of which of the two was more important was temporarily settled early in the Cold War; then, it was widely assumed, drawing on World War II experience, that there was *no defense* against bombers carrying nuclear weapons. With the belief that "the bomber will always get through," the US military prepared its offensive in the 1950s to destroy a ruinous variety of Soviet military and urban-industrial targets. Such an offensive was no longer sustainable by the mid-1960s. With increasing Soviet nuclear capabilities, the US realized that it could not prevent the Soviet Union from inflicting great harm on the US in any conceivable wartime scenario. Consequently, US government leaders abandoned the pretense that nuclear weapons could be used offensively to serve defensive purposes, that is, to protect the US from Soviet retaliation. But the US did not abandon offensive thinking. The US turned from offensive denial-based principles to the offensive punishment-based principles of AD doctrine.

16

The emphasis in US policy tilted a bit more toward defense in the late 1960s with increased US interest in anti-ballistic missile (ABM) technology (missiles that could intercept other missiles) and the principles of a defensive, denial-based strategy. That a strong US defense could lessen the burden for US offenses by shielding the US from attack was not a winning argument, though, given the limited defensive technologies available at the time and formidable Soviet (and emerging Chinese) offensive capabilities. The 1972 Anti-Ballistic Missile Treaty, which sharply limited defensive deployments, momentarily ended the policy debate over the value of a defensive-denial-based posture. It did not end the debate, however, over the wisdom of US reliance upon the offensive-punishment-based strategy of AD. By the 1970s, offensive-denial principles came into favor again with an increased interest in (war-fighting) counter-force options—using US missiles to destroy Soviet missiles before they were launched. The idea that US forces could engage in "damage limitation" (a policy rejected earlier by the Kennedy administration)—reducing US damage by destroying Soviet military forces—was a principal focus of Carter and Reagan-era policy. The defense gained some ground in the Reagan years with the denial-based "Strategic Defense Initiative," which envisioned protecting the US from attack with an ambitious array of defensive technologies. Through its offensive and defensive efforts, the Reagan administration clearly registered its distaste for the (offensive-punishment) principles of AD that left the US "hostage" to Soviet intentions.

During the Cold War, then, offensive logic generally prevailed, but not always obviously given the variety of policy participants and changing theoretical fashions. In comparison, contemporary US policymakers were quick to embrace offenses *and* defenses, and to discount each as a substitute for the other. For example, the US is building a defense against a rogue-state missile attack despite the continued effectiveness of US retaliatory threats and the deficiencies and costs of a missile defense. Similarly, the US touts the advantages of going on the offensive against terror groups when the US will enjoy some home-turf advantages with a defensive approach, without the disadvantages of an offensive approach. These disadvantages include trying to engage the enemy on too many fronts and aggravating rather than solving the terror problem.

Pivotal assumptions: rationality and non-rationality

Cold War-era deterrence theory was developed and applied with the assumption that the US and Soviet Union were rational, as judged by whether their behavior was appropriate given their goals. Thus, when increased Soviet retaliatory capability created a nuclear stand-off with the US in the late 1960s, the prevailing view was that US–Soviet relations would become more stable. The US and Soviet Union would shy from provoking each other given their *rational* recognition that things could get out of hand and that a nuclear exchange would bring prohibitive costs and serve no useful purpose. Indeed, the rationality assumption was embraced across the policy spectrum. "Dovish" critics of AD doctrine who argued for a "minimum deterrent force" had faith that the Soviet Union would not welcome retaliation from the US if it possessed but a small, invulnerable nuclear retaliatory force. US war-fighting proponents based their criticism of AD doctrine upon the assumption that retaliation could not be executed rationally because it was tantamount to suicide. The war-fighting solution was to make US retaliation "rational" by improving the accuracy and responsiveness of US forces, reducing the vulnerability of US weapons and/or cities, and developing the capability to respond to Soviet attacks at whatever level they occurred. A faith in rationality also influenced the thinking of policymakers who sought a middle course between denial and punishment—those who shared the basic optimism of AD proponents yet desired some military options, for coercive effect, should a nuclear war breakout. In their view, nuclear attacks by the US on peripheral targets could send clear messages to the Soviet Union about US resolve, to facilitate bargaining.[1] Presumably, the Soviet Union was rational enough to read these messages as intended, which could lead to de-escalation.

Again, policymakers were not always true to their abstract assumptions. By planning to attack the Soviet leadership directly through so-called decapitation attacks or by planning to destroy Soviet command and control centers to weaken the Soviet capability to fight a protracted nuclear war, policymakers were effectively limiting the Soviet ability to engineer a "rational" response—in fact, increasing the chances of a preprogrammed and spasmodic Soviet response. A Soviet Union, decapitated and crippled from US strikes, could hardly participate in a bargaining strategy to impede escalation to all-out war. If it could, it might not want to negotiate inasmuch as a "headless" Soviet Union might have little left to "discuss." Indeed, facing its supposed "worst-case" scenario—a loss of political power—the surviving Soviet leadership might well be inclined to do what was otherwise irrational—launch an all-out attack on US cities.

Logical contradictions aside, doctrinal advocates defined problems and sought solutions in the Cold War period by harnessing the rationality assumption; and, if anything, the perceived dangers of that period stemmed from the assumption of a "hyper-rational" adversary that would adopt audacious—indeed, catastrophically reckless—tactics with the sublime conviction of having assumed but a "calculated risk." The rationality assumption, which was once accepted so easily by hawkish AD critics, is now rejected out of hand by those who define the "rogue-state" problem. Whereas critics had once supposed the Soviet Union to be coldly rational, US policymakers now routinely suppose that the adversarial leadership—given its commitment to ethereal objectives, mercurial temperament, opaque reasoning, and obliviousness to cost—at best, borders on the irrational. On this point, the President's *National Security Strategy* (Bush 2002a: 15) warned that "we cannot let our enemies strike first" because "deterrence based only upon the threat of retaliation is less likely to work against leaders of rogue states more willing to take risks, gambling with the lives of their people, and the wealth of their nations." From this policy perspective, the US is left with no choice but to assume the worst and to take extreme measures now rather than allow the adversary to strengthen its capabilities for some future attack.

The reality, though, is that despotic leaders such as Saddam Hussein have acquired and held power—in the case of the former Iraqi leader, over turbulent decades—through cold-blooded, cost-benefit analyses, a willingness to place personal interest above all else, and a strong instinct for survival. Consequently, the portrait of these leaders etched by their behavior does not match the caricature promoted in US policy circles. What can be said of the logical faculties of a leader who willingly sacrifices *everything* purely for the satisfaction of destroying another? It is one thing to make the supreme sacrifice to leave the world a "better" place; it is quite another to do so out of a spiteful impulse to punish the US for its democracy or policies. Even the al-Qaeda leadership does not pursue destruction for its own sake.

Popular lore, and the limited imagination of Hollywood film-makers, certainly feed a fixation with the "madman"—the mythical mad scientist who works through his personal traumas by taking revenge on society with a dastardly deed that outshines prior dastardly deeds. But irrationality is not akin to recklessness and impulsiveness. Irrationality can lead to extremely conservative behavior—an unwillingness to take even those risks that might be necessary for self-preservation. People might seek to avoid surgery even if the chances of surviving an ailment without surgery are poor. People actually avoid going to the doctor out of fear that they will receive bad news of an illness that is life-threatening, though potentially curable if treated in time. Moreover, people are victims of psychological maladies *by degree*. Although people can be judged either legally competent or legally incompetent by the courts, the reality is that some people are *more* or *less* rational than others and that even the incompetent can exhibit rationality. Most psychotics do not wander out into traffic or jump from bridges because that is the easiest way to a destination. Thus, even those who appear disconnected from reality and logically impaired can defer to considerations of cost. In other words, deterrence does not require perfect rationality nor even logical thinking to work (Jervis 1979: 299).

These conclusions have implications for a US effort to deter rogue states—and terrorists, too. That terrorists consider the costs of acting is a basis for assuming that they can be deterred *under some circumstances*. On this point, there is evidence, for instance, that some terrorists will not act if their lives are placed at risk. After all, the leaders of terror groups display their survival instincts (and refine their survival skills) through years of imprisonment, harsh treatment in confinement, and/or efforts by governments and rivals to outmaneuver, capture, or kill them. In any case, it is a rare leader who can view their organization's interest apart from their own. Having created the organization in their own image, or struggled to put their own imprint on the organization, it is easy for leaders to reject self-sacrifice as a useful option.

Most certainly, the danger posed by al-Qaeda and other fundamentalist Islamic groups is that their grandiose aspirations require and legitimize actions of unlimited magnitude and scope. Religion-inspired terrorists "perceive violence to be a sacramental act, or divine duty, executed in direct response to some theological demand or imperative" and are unconstrained by norms and political exigencies: "they are at once activists and constituents, who execute their acts for no one but themselves" (Gurr and Cole 2000: 30). But a strong commitment to grand objectives is also reason for restraint: even zealous operatives exercise caution when their capture or death will hinder the successful completion of an important mission. To the extent, then, that terrorists act furtively, meticulously plan attacks, devise big and complex operations, and invest in the future, they convey and/or create a susceptibility to costs that creates opportunities for deterrence.

In the final analysis, policymakers have little choice but to assume that rationality informs the behavior of rogue states and terrorist groups. The alternative is to assume that US adversaries are unpredictable and to surrender any hope of prudence in devising a US response. True, there are clear advantages in "worst-case" planning. At one level, it makes sense to acknowledge, for instance, that North Korea can turn over its nuclear weapons to terrorists or plant a nuclear weapon within the continental United States for possible use. But assuming that whatever could occur has occurred (or will occur) is unreasonable—and dangerous—for it justifies unrestrained action against North Korea, Iran, and other countries harboring nuclear ambitions, with consequences that the US and its allies can ill-afford. It is worth remembering that restraint prevailed in US (and Soviet) Cold War-era policies that were reputedly constructed around "worst-case" thinking.

Extending deterrence: the challenges to retaliatory policies

During the Cold War, the challenges of "extended deterrence" were captured in a nagging question posed by the French leadership: when push came to shove, "would the US trade New York for Paris?" The question was a reasonable one—indeed, an inevitable one—given the horrific consequences that could befall a country that aids another under nuclear attack. It is debatable whether the US would have stood by its commitment to retaliate massively with nuclear forces for a Soviet nuclear attack on the US. It is even more questionable whether the US would have willingly accepted catastrophe to stand by commitments in other parts of the world.

In a critical sense, the issue of extended deterrence was an extension—albeit a magnification—of the dilemmas that inspired war-fighters to lament the pitfalls of US AD doctrine. With extended deterrence, US policymakers were forced once again to confront the limits of their deterrence logic. It is one thing for the US to invite incineration when the US had already incurred significant destruction as a result of a Soviet strike: even a "surgical" Soviet attack on US nuclear targets could well have left millions of Americans dead. It is another thing to suppose that the US would view these levels of destruction as the acceptable price for standing by a friend. Thus, US efforts to craft doctrines and procure and deploy forces for fighting "local wars" or a "flexible response" to Soviet transgressions in Europe had to concern US allies. If the US was serious about using nuclear weapons in defense of US national interests, which were publicly defined (in the North Atlantic Treaty) to include Western Europe, could not US efforts to limit fighting to Europe or to acquire conventional and nuclear options for a flexible response short of an all-out nuclear war be taken as a sign that the US was unwilling to do *everything* required to thwart Soviet objectives?

The challenges of extending deterrence to US allies lessened but did not subside entirely with the end of the Cold War. North Korea's territorial ambitions on the Korean peninsula remain a justifiable source of US concern, as do Chinese claims to Taiwan; and Iran could conceivably use its nuclear power to change political realities in the Middle East: some Iranian leaders have promised as much, and their anti-Israel rhetoric is cause for concern. Indeed, there is reason to believe that rogue states will attack vulnerable US allies to punish the US, and even terrorists present the US with an extended deterrence challenge when picking their targets to reduce the likelihood of a US defensive or retaliatory response.

Extended deterrence problems have also arisen in new forms. The superpowers had managed Cold War arms control within a framework in which deterrence was the intent of the parties. Thus, the superpowers cooperated with each other to create institutional arrangements and procedures that permitted verification of each other's capabilities. Such cooperation was made possible because the danger from "hidden" capabilities was slight: with large, robust nuclear arsenals, neither superpower could quickly gain the upper hand by exploiting some new technology or weapons in a secret arsenal. Now, the US has the difficult task of dissuading countries from acquiring weapons (and dangerous technologies), and others from supplying them, when the recipients have strong incentives to acquire weapons that even in small quantities promise their possessors large political and military benefits.

Nevertheless, the challenges of extending deterrence can be overstated. Although the US is unable to stop all attacks on US allies or to forestall the transfer to third parties of dangerous technologies for possible use against the US or its interests, the US is positioned to deter attacks on third parties by states that have little to gain and much to lose from those attacks. Similarly, the US is positioned to deter attacks by terrorists on less-defended targets when the costs and risks of attacking these targets exceed their value to the attacker. In this sense, the problem of extended deterrence is an aspect of a broader deterrence problem, which permits some optimism about whether and when challenges to the peace will occur.

Compellence: the limits of deterrence

Cold War-era theorists recognized that the logic of deterrence could not be extended to include problems of "compellence." Whereas deterrence involves the use of coercion to prevent something from happening, compellence involves the use of coercion to cause something to happen (Freedman 2004: 109–12; Schaub 1998).

That deterrence is understood to involve attempts by a party to *preserve* the status quo means that deterrence messages are easier to convey than compellence messages are (Schelling 1966). The deployment of North Atlantic Treaty Organization (NATO) forces along the borders of Eastern Europe, and of coalition forces along the Kuwaiti border with Saudi Arabia in 1990–1, sent an unmistakable signal to the Soviet Union and Iraq, respectively, that a confrontation would ensue if either country were to cross a line. In contrast, compellence signals are ambiguous. Deeds, and even words, can inaccurately convey the intentions of the sender—including that its aim is compellence. When the US and its allies sought to compel an Iraqi withdrawal from Kuwait, the Iraqi leadership could reasonably ask whether a partial or staggered Iraqi withdrawal would satisfy US demands and whether the US meant to back up its demand with force. After all, coalition forces arrived on the scene after the damage had been done (Iraq had assumed control of Kuwait), trickled into the region over a six-month period that suggested a lack of urgency, and deployed arguably in defensive positions on Saudi territory. Even repeated US warnings, backed by UN resolutions, could have signaled to Iraq that the US preferred to avoid a fight. As a result, Iraq might have interpreted the US compellence message as a deterrence message and believed, then, that Iraqi forces were safe as long as they did not move beyond Kuwait.

Apart from these signaling problems, compellence is difficult to achieve. No one wants to concede what they actively sought; obviously, they valued it enough to have pursued it in the first place. This is true whether the thing of value is territory, dignity, or political prerogatives. There is a psychological aspect to this, too. Decisional studies indicate that people are more willing to concede opportunities for gain than to accept a loss—that people are more attached psychologically to what they possess than what they do not (yet) possess (Kahneman and Tversky 1979; Levy 1997; McDermott 2001). The acceptance of loss is made harder still when compellence requires "visible" concessions that cause the conceding party to lose face or incur reputation damage. These same costs are not incurred when the intent of a party is deterrence. It is easy for leaders to justify their inaction, when encountering a deterrent threat, by claiming that they had no intention of challenging the status quo in the first place. Rhetorically, at least, all states value peace and respect the sovereignty of their neighbors.

Compellence issues will confound US policymakers as they confront new threats. Much of the contemporary nuclear debate, applied to fledgling nuclear powers, is really about compellence. The US, in pressing its case against Iran and North Korea, is effectively asking them to give up something, that is, to concede their present or future nuclear capabilities and, thus, to accept their increased vulnerability to an attack (possibly from the US). The challenges for US policymakers are that much greater when compellence issues combine with deterrence issues and which of the two is most critical is not well understood. Some of these challenges occur when compellence policies have deterrence effects, and vice versa. Through current efforts to *compel* North Korea and Iran to dismantle their nuclear infrastructures or at least open them to outside inspections, the US is seeking to send a coercive message to *deter* other states from pursuing illicit weapons programs. A problematic reverse relationship is also apparent. By inflicting political and economic costs on Iran, for example, to *deter* it from acquiring nuclear weapons, the US seeks to *compel* Iran to drop its plans to build these weapons. If Iran were deterred from conducting nuclear activities in secret or from acquiring technology needed to produce a nuclear bomb, the added costs and risks of pursuing weapons under these new conditions could hurt the case within Iran for procuring the bomb. Other challenges occur when the issue differs from the adversary's perspective. Compellence is involved when the US seeks to "deter" actions by governments that perceive their leadership, credibility, or national prerogatives at stake. From their perspective, these governments are being *compelled* to end their support for terror groups. This is a costly concession for governments that are committed by reason of ideology, economic dependence, or political weakness to support these groups. The costs for these governments include a loss in domestic support, regional stature, and (perhaps) standing within the Islamic world. Compellence is involved, as well, in efforts that directly target terrorist groups. The problem of deterring terrorists effectively becomes a challenging compellence problem when, for these groups, not acting is equivalent to surrendering their movement, popular following, and *raison d'être*.[2]

23

A key to successful policy in the contemporary age then is knowing when deterrence principles apply and when they do not. Crucial, too, is being able to recognize how deterrence and compellence problems overlap (in the near or long-term and in the vicinity of a conflict or around the globe) to complicate policy problems—but, also, to create opportunities for policymakers who can adjust their goals when they prove unrealistic. Policymakers might recognize, for example, that compellence instruments aimed at stopping nuclear proliferation or ending state support for terrorism promise some success—from a deterrence perspective—if simply constraining the growth of the North Korean nuclear force, delaying the Iranian acquisition of nuclear weapons, forestalling the hand-off of nonconventional weapons by states to terrorist groups, or preventing a terrorist attack on the US.

Conclusions

The Cold War is over. Yet, new challenges invite fresh thinking about Cold War-era policy dilemmas and their solutions. Put simply, the methods and issues of the Cold War period *are* relevant today. This is true whether the issue is the relative contribution of offense, defense, punishment, and denial; the implications of assuming that US adversaries are rational; the ability to extend deterrence to various areas of the world; or the challenges of deterrence versus compellence. It is also true whether US adversaries happen to be small states exceptionally positioned to harm the US or terrorist groups positioned to move surreptitiously across national borders, insinuate themselves into society, and pursue nefarious goals.

Figure 2.1 juxtaposes the principles that supported the US doctrine of AD and its main Cold War competition in pure war-fighting doctrine. Clear from the figure is that AD focuses policy in the upper-left, offensive-punishment cell while war-fighting shifts the focus to the bottom half of the table. Although the balance sought between the two columns varied with intellectual tastes, war-fighters tended to favor the offense (hence, the bolder print) and to view defense as subordinate, at best, to the offense. In accepting the denial-based reasoning of their Cold War-era predecessors, current US policymakers seek to bring the offense and defense into greater balance. Of course, their goal has not generally been to promote a deterrence strategy: unlike Cold War-era war-fighters, US policymakers do not express much optimism that deterrence can work against the main enemy—now, rogue states and terror groups. Admittedly, some US policymakers acknowledge the subordinate contribution of an offensive punishment-based strategy for combating the rogue-state threat. They recognize that even rogue-state leaders can be deterred by potential losses: their concern is what these leaders might do in *extraordinary circumstances*, as when their leadership is threatened. Some policymakers concede, too, that

	Offense	Defense
Punishment	**ASSURED DESTRUCTION**	
Denial	**WAR-FIGHTING**	War-fighting

Figure 2.1 Cold War-era nuclear deterrence doctrines

punishment could dissuade state supporters of terrorism and the less dedicated members of the terror network.

As will become apparent, all three cells can contribute to sound US strategies to counter rogue states and terror groups. The following chapters show that US policymakers *should* retain the offensive punishment-based focus of AD doctrine to deter rogue states. The alternative, a denial-based strategy, remains problematic for the US despite the relatively modest size of rogue-state arsenals. Preventative and preemptive strike options (discussed in Chapter 3) against rogue states cannot guarantee success and (along with a strategy of missile defense) invite a host of confounding problems. In addition, this book establishes that US policymakers *should* look primarily to a defensive denial-based strategy (the bottom-right cell) to address the global terror threat. Although no strategy will forestall all terror attacks on US and non-US targets, a deterrence strategy based on defensive principles can help the US *manage* the terror threat without exacerbating it or dissipating scarce US resources. In the final analysis, the US will be better off with a defense-led approach than with available alternatives. (These conclusions are presented in Figure 7.1 in the last chapter.)

3

ON THE OFFENSIVE AGAINST ROGUE STATES

Punishment and denial strategies

Preemption is among the counter-force options of a denial strategy. Its prominence in US strategy stems, in part, from the fear that attacks by rogue states cannot be deterred by retaliatory threats backed by the full weight of the US nuclear and conventional arsenals. Indeed, a prominent justification for the US invasion of Iraq in 2003 was that leaders, such as Saddam Hussein, cannot be dissuaded by threats of reprisal.

Branding Hussein and others as "rogue" leaders arguably brings needed attention to their heinous practices. The Iraqi leadership under Saddam Hussein directed the slaughter of Kurds and Shiites within Iraq and instigated a war with Iran that resulted in a million deaths. Hussein even used chemical weapons against the Kurds and Iran to exact a higher death toll. For its part, the North Korean regime has pursued policies that have bankrupted the North Korean economy, oppressed its people, and impeded a resolution of conflict on the Korean peninsula over half a century after the Korean armistice. The North Korean regime is troublesome, in fact, because of its willingness to tolerate a high degree of separation from the global community—"to endure international isolation and opprobrium" (Coe 2005: 76)—and to flaunt its violations of treaty commitments. For instance, in 2003, North Korea withdrew from the Nuclear Non-Proliferation Treaty (NPT) and, in the recent past, has claimed to possess a nuclear arsenal and program for enriching uranium, and conducted a nuclear test.

Whether or not the leaders of these states are unusually "roguish" in a world that is abundant in heinous regimes, the behavior of these leaders can still conform to deterrence principles. In actuality, the threat from leaders that—by merit of their impulsiveness, illogic, or animosity—simply cannot be deterred has dubious logical and factual foundations. The first section of this chapter assesses these foundations by examining the contemporary usefulness of a punishment strategy for addressing the rogue-state challenge. The second section assesses whether the US can obtain its objectives through the denial option of preemptive war—attacking an adversary when it is about to strike—and preventative war—attacking an adversary before it is able or willing to strike.[1] In this section, denial strategies are viewed, not as an articulation of (war-fighting) deterrence logic, but through the lens of contemporary US policy, as an alternative to relying upon deterrence.

In search of an intermediate path between dissuading nuclear-armed states from attacking and attacking them before they can use their weapons, this chapter concludes by examining the effectiveness of trade sanctions (blockades and embargo) as responses to proliferation. With sanctions, the issue is not how to prevent countries from *using* nonconventional weapons but how to prevent countries from *acquiring* or *retaining* these weapons. Consequently, appreciating the utility of sanctions requires that the proliferation problem be understood through: (a) a deterrence logic (that combines punishment and denial principles); and (b) the non-deterrence logics of *denying* states access to dangerous technology and of *compelling* states to concede their weapons programs.

Punishment: a counter-value strategy

Present fears about rogue-state leaders are an understandable response to the unsettling prospect that the US will face new or changing dangers. In this respect, the current US predicament is not all that different from the Cold War version. In the early Cold War years, US policymakers were uncertain that a deterrence relationship between the US and Soviet Union could be achieved; to the contrary, they feared that, once the Soviets possessed nuclear weapons that could reach the US, they would use them (Lebovic 1990: 74–5). Consequently, policymakers in the Eisenhower administration seriously considered initiating a preventative war with the Soviet Union (Sagan and Waltz 1995: 59–61). The same fear of the unknown surfaced in the 1960s when China emerged as a nuclear power. US policymakers worried that US preponderance, interests, and commitments would do little to deter China, which, unlike the Soviet Union, had not embraced existing rules of strategic competition. Such fears never fully subsided. Whereas AD proponents saw Soviet leaders as conservative and disinclined toward risk, others (e.g. war-fighters) depicted the Soviet Union as a dedicated, cold, and opportunistic nemesis, one that would strike the US if an opportunity presented itself (Nitze 1984/5; Pipes 1980, 1984). In the years leading to the collapse of the Soviet Union, Soviet goals were subjects of grave concern in the official policy statements of the US Defense Department, whose annual reports still referred to the Soviet quest for "world domination." Soviet motives also catalyzed debate within the intelligence community, in which the US defense establishment resisted evidence that the Soviet Union was fundamentally and irrevocably changing. For instance, reductions in Soviet military spending were dismissed by shifting the focus to the sheer size of the Soviet military force, unfailingly aggressive Soviet intentions, and limits to change imposed by the strangling influence of Soviet institutions. Indeed, evidence of declining Soviet military spending was read in dire terms when believed to forewarn of improvements in the quality of Soviet military capability, in that quantitative reductions could produce a "leaner and meaner" Soviet fighting machine. With the collapse of the Soviet state, "uncertainty" itself became a primary threat. US government officials warned that the former Soviet Union could revert to its former self even as they looked beyond the Soviet Union to discover the broader, less-defined threats associated with "regional conflict." By requiring that the US protect itself against potential challengers in all parts of the globe, the threat of the unknown seemed bigger and more dangerous than the Soviet Union ever was (Lebovic 1996: 23–5).

In the end, Morgan (2003: 29) puts it best when he notes:

> Many analysts now look back on Cold War era deterrence as qualitatively different because each superpower faced an opponent it understood, both lived with rules that provided a framework, and their long experience enabled each to understand the other's interests and perspectives. *This is mostly nonsense.* For the most part, the US did not understand the Soviet Union well at all, and vice versa.

Morgan (2003: 30) goes on to recognize that AD was a profound profession of ignorance. Policymakers disagreed considerably over what it would take to deter the Soviet Union. To the extent that the US defined AD with precision, it did so with widely varying estimates (in Department of Defense annual reports) that reflected the amount of destruction that the US believed it *could* efficiently inflict, not what the US *must* inflict in order to deter an attack. It was for lack of a better alternative, then, that policymakers (and arms control specialists) were guided by principles of a nuclear "balance" that could be taken to absurd lengths. Some war-fighters argued, for example, that the US had to mimic Soviet deployments—by relying upon heavy, and arguably vulnerable, land-based missiles—even if this would not give the US a usable military advantage. The idea was to offset a *hypothetical* Soviet belief that such deployments gave the Soviet Union an edge.

Given the poorly informed judgments of Soviet behavior in the Cold War period—despite active US monitoring of developments within the Kremlin, decades of historical experience upon which to draw, and a common cultural heritage—it appears reasonable to conclude that US policymakers are even less informed now about the values and beliefs of their adversarial counterparts, and that they will see *that* as a threat. In this vein, a former Bush administration defense official (Payne 2005: 139–41) warned of an enemy in pursuit of "some intangible, possibly unimaginable goal" that cannot be deterred by any threat to its "material values"—presumably its people, cities, or military capabilities.[2] For other defense officials, the threat lay in the possible irrational impulses of known US adversaries. Richard Perle expressed that view colorfully when, early in the first Bush term, he remarked before a congressional committee, "I frankly don't want to count on the rational judgment of a man [Saddam Hussein] who used poison gas against his own people" (Loeb 2001).[3] In actuality, it matters little whether US enemies are barbarously hateful and vengeful, or ruthlessly insensitive to cost. All such enemies will exhibit the same behavioral tendencies—a wanton indifference to behavioral consequences that brings the rationality of these adversaries into question. In impugning the rationality of US adversaries (or accentuating their vengefulness or cost indifference), policymakers mischaracterize adversary reasoning.

Iraq under Saddam Hussein

That the Saddam Hussein's of the world are irrational or effectively self-destructive is open to dispute. In serious policy analysis, Hussein was typically viewed as an audaciously strategic leader whose missteps reflect flawed judgments, but not illogical nor suicidal tendencies. Even Iraq's failed effort in Kuwait is attributable to the absence of a concerted US effort to deter him, Hussein's misjudgments about the likely response to his military actions from friends and adversaries, and his belief that the costs of doing nothing were prohibitive given threats to his leadership from abroad, inadequate economic support to Iraq from neighboring Gulf states, and the economic strains on Iraq from years of fighting a brutal and costly war with Iran (Stein 1992: 177–8). Moreover, in the years following the first Gulf War, Hussein tested the US commitment and escalated and de-escalated the conflict to affect the perceptions and resolve of his adversaries, but he did not throw his forces into battle to fight for a losing cause. Nor did he use his missiles incautiously. With the hope of undermining the anti-Iraq coalition, Iraq's attacks on Israel were carefully calculated to bring Israel into the conflict (Freedman and Karsh 1993: 436). During the war, Iraq refrained from using biological or chemical warheads; and, with the end of hostilities, it ceased missile attacks on Israel and Saudi Arabia.

Of course, the conflict between Iraq, its neighbors, and the outside world did not subside and continued for more than a decade, in which Iraq remained internationally isolated. In the end, Hussein invited his own demise through his maddeningly obstreperousness in preventing international inspectors from acquiring information about Iraq's weapons programs. His actions helped the US government to make the case (at least domestically) for regime change in Iraq. But Iraq's behavior was not inherently reckless. Instead, it was calculating and self-interested. It reflected Hussein's views that the disadvantages of cooperating with inspections outweighed the advantages (as discussed below) and his underestimation of US military strength and resolve (Woods *et al.* 2006). His fundamental tendencies did not change with events. When Hussein belatedly recognized his miscalculation, he sought to avoid a showdown with the US by employing foreign intermediaries to forestall a US attack; with the onset of war, he stuck to war plans that he believed would keep him in power (Wood *et al.* 2006); and, when Baghdad ultimately fell, he opted for self-preservation. He did not move to the front lines to command the loyal resistance in a counterattack against occupying US forces. His new lifestyle was optimized for secrecy, and self-protection. When finally captured, he capitulated without a fight, seemingly relieved to be rescued from the stress and deprivation of life on the run. To the very end, Hussein appeared to be serving no cause larger than himself. As Gaddis (2004: 91) observes, Hussein showed that he, like other such tyrants, was more into "survival than suicide" and "palaces than caves."

That Hussein would have paid a high price in Iraqi and non-Iraqi lives is no more sustainable an argument unless it supposes that he might have gained from doing so. But what would a nuclear attack on the US offer an Iraqi leader other than US civilian deaths? And why would Hussein care about US deaths if, as deterrence skeptics claimed, he viewed human life as cheap? Did Hussein care so much about killing Americans that he would willingly sacrifice scores of Iraqis for every American killed, perhaps risking his own life, his family, and, indeed, his (opulent) lifestyle in the process? Even war-fighters did not assert that Soviet leaders wanted to murder US citizens per se, let alone kill them in the "paltry" numbers that are within reach of rogue states: in fact, some war-fighters maintained that the US could "win" wars with US casualty rates considerably higher than those that rogue states could inflict on the US in the foreseeable future. By failing to indicate what an adversary could gain from a nonconventional weapons attack on the US, then, deterrence skeptics feed suspicions that the adversary has nothing to gain from an attack and, therefore, that a nuclear attack by a rogue adversary would be an irrational or self-destructive act. The US could certainly find itself under threat from an irrational or self-destructive leader, but history does not suggest that the probability is high or the implications are entirely bleak. After all, even the irrational can consider costs.

With Hussein languishing in an Iraqi prison, it might not seem useful to point to him as the model leader that so concerns US policymakers. But Hussein provides an important test case because he was heralded before his fall from power as the archetype of the new leader—a veritable "poster boy" for leaders that are implacably and irresponsibly hostile to US interests. Thus, it is appropriate to ask what has since been learned from his behavior. First and foremost, it appears as if Iraq could, in principle, have been deterred—to a degree that was unappreciated even by those who made the case against a US invasion (Mearsheimer and Walt 2003). He had chosen to destroy his stockpiles of prohibited weapons and to dismantle the Iraqi infrastructure for research, development, and production of nonconventional weaponry in a way that impaired an easy reconstitution of those capabilities.

The obvious question, then, is, "why didn't Hussein cooperate with outside inspections if he had nothing to hide?" One answer is a somewhat non-rational one: "this is what despots do; they hide things." Their instincts are inherently nondemocratic. A rational variant of this explanation is given weight by evidence obtained from post-war interrogations of Hussein's advisors. From Hussein's standpoint, it seems, revelations that would establish Iraqi *compliance* with international demands were actually damaging to Iraq if revealing its prior noncompliance or deception, or if arousing suspicions that would lead inevitably to more international demands for Iraqi disclosures and access.[4] Iraq had good reason to fear that its actions would be interpreted in the most negative light. Even Iraq's efforts to adhere to the disarmament requirements of UN resolutions were interpreted unfavorably by US intelligence as efforts to circumvent these resolutions. In one instance, communication intercepts of an order to destroy chemical agents were treated as confirmation that Iraq was retaining these weapons (Woods *et al.* 2006). For a dictator such as Hussein, ordering documentation and equipment to be destroyed, operating surreptitiously, and making facilities and program personnel unavailable to inspectors—that is, stonewalling and obfuscating rather than cooperating—was the logical path.[5]

But this response only begs the question of why the Iraqi strategy was incomplete. What good is disarming if those who demand it remain unsatisfied? A second interpretation of Iraq's behavior is parsimonious in comparison, and non-contradictory. Hussein was pursuing a rational deterrence strategy based on "strategic ambiguity." On the one hand, he sought to deprive the US of a "smoking gun" that could be used as a pretext for military action against Iraq. On the other hand, he could not offer definitive proof that he had disarmed (as Libya did through the public destruction of weapons and the opening of governmental records and complexes to inspection) because this would leave Hussein without a credible threat to employ exceptional weapons to ensure the survival of his regime in a hostile international environment.[6] This interpretation has also been substantiated by post-war revelations. Hussein purportedly told his Revolutionary Command Council that Iraq did not have WMD but that to admit this publicly would invite an attack (Woods *et al.* 2006). As one US intelligence source put it before the US invasion, "Iraq would never admit that it did *not* have WMD because it would be tantamount to suicide in the Middle East."[7] These conclusions would later be echoed after the 2003 invasion by two successive heads of the Iraq Survey Group—the US postwar investigation team on Iraqi WMD (Duelfer 2004; Ellis and Kiefer 2004: xvii). From Hussein's standpoint, then, Iraq's interests would be compromised through disclosure that he did *or* that he did not possess nonconventional weapons. He thus chose the perilous middle course.

Iraq artfully employed its failed deception strategy. Following the strategy, Iraq could manipulate events and disclosures to reinforce the credibility of the Iraqi deterrent threat. Its tactics toward UN inspections—refusals to cooperate, suspicious foot-dragging, sporadic cooperation, and reneging on agreements—created the useful appearance that Iraq had something to hide. The strategy even offered plausible applications when the US invasion came. Iraq could create the appearance that it was preparing to use its "arsenal" in a final showdown by maintaining chemical defense battalions, or even planting protective chemical suits and antidotes with Iraqi troops in the approach to Baghdad. Ironically, then, a leader who allegedly defied the rules of behavior within the international community might actually have refined the deterrence strategy that had been a mainstay of the community.

Even the Bush administration conceded implicitly, in its wartime and post-war public comments, that Hussein was deterrable. When the administration voiced concerns that Hussein would use chemical and biological weapons in a final defense of Baghdad, it undermined the logic that had been used to justify the invasion. It now suggested that Hussein would take the ultimate step when absolutely necessary, not when US targets of opportunity first presented themselves. Then, when Iraqi nonconventional weapons proved elusive, the administration resorted implicitly to deterrence logic by maintaining that Iraq had hidden its weapons. It thus implied that Iraq was somehow deterred by the prospect of political or economic costs should those weapons be discovered—costs that outweighed the benefits that Iraq would accrue from employing or threatening to employ those weapons. For that matter, the administration implied that these costs loomed so large for Hussein that he would disarm himself by sidelining those weapons when he most needed them—when facing his alleged worst-case scenario, an imminent collapse of his regime.

Iran and North Korea: nuclear programs

The US defeat of the Saddam Hussein regime still left in power two regimes of concern to the Bush administration. Unlike Iraq under Hussein, both Iran and North Korea have robust nuclear programs that, at the very least, could produce nuclear weapons at some point in the future. Indeed, the nuclear program in North Korea is well advanced, and North Korea probably possesses a nuclear arsenal, as suggested by the country's nuclear test in 2006. In the Iranian case, the evidence suggests only that Iranian technologies *could be* diverted to construct a nuclear weapon. An Iranian nuclear program is not itself an indication of illicit intent in that Iranian oil is a depletable commodity and Iran has rising energy needs (Howard 2004:

98); and countries can have noneconomic reasons for pursuing nuclear power. Even the furtiveness of the Iranian nuclear program could be, as Iran claims, an attempt to protect it against possible attack by the US or Israel or else, following in the footsteps of the illusionary Iraqi program, a false front. To quote a high-ranking intelligence official within the US State Department, "The Iranians don't necessarily have to have a successful nuclear program in order to have deterrent value," "they merely have to convince us, others and their neighbors that they do" (Jehl and Schmitt 2005). The reality is that outsiders have incomplete knowledge about the nuclear programs and intentions of both countries—and much of what is known comes from communications intercepts, satellite imagery, and international inspectors and certainly not the detailed revelations of program insiders, as available post-mortem on the Iraqi program. Since 1995, US officials had estimated that Iran was as close as five-years away from being able to produce a nuclear weapon, but an authoritative review, in the form of a National Intelligence Estimate in 2005, predicted that Iran was ten-years away from producing enough uranium for a nuclear device (Cordesman 2004: 125–38; Linzer 2005a). The relevant questions, then, are whether Iran and North Korea will pursue the nuclear option and, if they do, whether the intentions and behavior of these countries suggest that they can be deterred from using their weapons. Some perspective on the answers can be obtained by examining the status of these countries' nuclear programs, their justifications, and the bargaining tactics with which these countries have pursued their programmatic goals.

Status

Iran is not yet believed to have reached the point of technological self-sufficiency, and even the ten-year estimate was predicated upon the assumption that the program would proceed uninterrupted by political or technological developments, such as Iran's recent shutdown of uranium enrichment to facilitate negotiations with Europe over the future of the Iranian program. Iran is in the process of constructing nuclear reactors, centrifuge facilities for uranium enrichment, and conversion plants that can transform (yellowcake) uranium ore into hexaflouride (UF6) gas which can be enriched (by centrifuge) to produce fuel for nuclear power plants (or, with further enrichment, weapons-grade uranium). To produce nuclear weapons, Iran must complete its nuclear reactors and develop a spent-fuel reprocessing capability or expand its efforts to build centrifuges to enrich uranium. Should Iran develop these capabilities, it would still need to transform the materials into weapons. Iran has invested considerable resources in missile development, through cooperation with North Korea and Russia, and can deliver missiles over intermediate ranges (which allows Iran to hit Israel, US forces in the Gulf, and parts of Europe). Indeed, some experts maintain that a missile investment makes sense for Iran only if it plans to deliver nonconventional warheads (Howard 2004: 101). Their argument is bolstered by evidence that Iran has flight tested a missile with an advanced nose cone with a payload space that is too small arguably to offer much "bang" with conventional explosives (Broad and Sanger 2005a).

Of concern, then, is the belated and incomplete disclosures of information by Iranian officials on their nuclear program. Iran confessed the existence of nuclear facilities only after evidence of their existence was obtained from other sources; and Iran surprised UN inspectors in 2003 with its advanced and extensive investment in centrifuges that could enrich uranium. Disquieting disclosures by Iran of its additional nuclear facilities, foreign acquisitions of nuclear material, and conversion tests on nuclear material fed suspicions further that Iran was seeking to acquire nuclear weapons, as did ongoing Iranian efforts to block access by UN inspectors to suspicious sites (even after Iran signed—though it did not ratify—the Additional Protocol of the NPT in December 2003, which permitted international inspectors to launch intrusive and wide-ranging inspections).[8] Thus, outside analysts disagree over whether Iran is simply trying to keep its options open and has not yet decided to pursue nuclear weapons.

North Korea's nuclear infrastructure now encompasses the complete nuclear fuel cycle. Thus, North Korea is believed self-sufficient in the production of the full range of materials required to construct and deliver a nuclear weapon. North Korea possesses nuclear reactors, reprocessing facilities for spent fuel, and indigenous supplies of uranium; it is believed to have imported centrifuges for producing highly enriched uranium (HEU); and it possesses an impressive missile delivery capability (though likely insufficient, at this point, to deliver nuclear weapons against targets in the continental United States). The US Central Intelligence Agency (CIA) estimates that North Korea already has diverted sufficient plutonium from its reactor program to build one to two nuclear bombs and might have adequate plutonium for four to six additional weapons if it had reprocessed 8,000 spent fuel rods removed from temporary storage ponds, and even more weapons with the shutdown of its 5-MW nuclear reactor in 2005, which allowed North Korea to extract additional fuel rods for reprocessing (NTI 2006b). Whether these materials have been assembled into weapons or paired with North Korean missiles is another matter.

Under the 1994 Geneva Agreed Framework (negotiated by the Clinton administration), North Korea was bound to cap its production of plutonium, accept certain safeguards to inhibit the diversion and reprocessing of nuclear materials for weaponry, account eventually for all of its plutonium, and dismantle its plutonium-production capabilities. Progress was to be pegged to benefits that included the supplying of light-water nuclear reactors (which are a low nuclear proliferation threat) and heavy oil to North Korea. Despite its pledge, North Korea hindered international access to its suspect nuclear facilities and apparently acquired gas-centrifuge uranium enrichment technology from abroad. The issue came to a head in 2002–3, when North Korea repudiated various international agreements and made compromising disclosures. Indeed, a North Korea delegate to trilateral talks in Beijing supposedly took the US Assistant Secretary of State aside to announce that North Korea "[had] nuclear weapons, [would] not dismantle them, and might transfer or demonstrate them" (James A. Kelly quoted in Coe 2005: 73)—a claim that North Korea echoed in the years to follow (NTI 2006b). Nonetheless, questions remain about the size of the North Korean nuclear arsenal, how many additional weapons North Korea can and will produce, and whether North Korea is ultimately committed to its nuclear weapons program and can be convinced to abandon it. Outside analysts are not in agreement over whether Kim Jong-il will renounce the nuclear option with the "right deal" (though most are pessimistic).

Justification

The idea of a nuclear-armed Iran or North Korea is an unappealing prospect for the US and its allies. But the mere pursuit of such weaponry by these or any other country is not by itself an indication of malevolent intent. States have sought (and rejected) nuclear weapons for a variety of domestic and international reasons. The irony is that weapons that are viewed ominously by potential target countries might not be pursued by their potential attackers primarily for use, compellence, or even deterrence.[9]

Leaders have been impressed by hypothetical benefits that accrue from possessing nuclear weapons. These weapons place their possessors within an exclusive club composed largely of global powers. Hence, these weapons provide a cheap and (relatively) quick means by which countries can acquire the trappings of global power status and overcome constraints imposed by resource, technological, and economic limitations. At the same time, these weapons can bolster the domestic positions of leaders who, in giving their country a "special" weapon, can appeal to nationalistic sentiment and feed a general sense of security. People are bound to feel more protected with the "ultimate" weapon than without it, especially when adversaries possess or might possess it. That the acquisition of nuclear weapons could have deleterious second-order effects—if triggering an arms race or a disarming, preemptive attack—is an argument that is lost on a public overcome with the pride of ownership. Thus, added to the somewhat predictable support these weapons enjoy from scientific laboratories, the nuclear power industry, and the military is the support that these weapons receive from broad domestic coalitions of ethnic, religious, nationalistic, cultural, and economic interests "backing an inward-looking strategy" (Solingen 1994: 140). Nuclear weapons are used symbolically by this coalition—in nationalist or subgroup appeals—to strengthen its domestic support (Solingen 2001: 379). Government leaders might have insufficient political strength or will to oppose these interests; much depends, it has been argued, on the offsetting strength of a broad liberal-izing coalition that includes business and professional groups. These groups view these weapons as obstacles to the integration of the nation within the global community, fear the effects of political and economic retribution should the nation acquire "nuclear pariah" status, and/or do not welcome the expanding governmental role and prerogatives that nuclear programs surely bring (Solingen 1994: 139).[10] In the end, "nuclear weapons, like other weapons, are more than tools of national security; they are political objects of considerable importance in domestic debates and internal bureau-cratic struggles and can also serve as international normative symbols of modernity and identity" (Sagan 1996/7: 55).

It is against this backdrop of competing internal forces that nuclear developments in Iran and (perhaps) North Korea must be understood. These forces can strengthen the domestic appeal of policy assumptions that complement a deterrence logic.

In Iran, conservative clerics, with the upper hand on matters of national security, are pushing for a nuclear-armed Iran: "the ideologues, who view a conflict with the United States as inevitable, believe that the only way to ensure the survival of the Islamic Republic—and its ideals—is to equip it with an independent nuclear capability" (Pollack and Takeyh 2005). Central to the clerics "is the notion that the Islamic Republic is in constant danger from predatory external forces, necessitating military self-reliance" (Takeyh 2004/5: 55). Conservatives have been highly successful in making their case to the public by framing the issue in terms of a right to acquire nuclear technology and thus a matter of national determination, sover-eignty, and dignity; resistance to foreign interventionism; and technological development. By alluding to US efforts to deny Iran access to any nuclear technology, Iranian leaders can argue domestically that US efforts are part of the larger plan to "keep Iran backward and dependent" (Chubin and Litwak 2003: 105–6). Such arguments have also shaped Iran's international pitch, as when Iran's nuclear negotiator observed sarcastically, "second-class countries are allowed to produce only tomato paste" (Broad and Sanger 2005a). The conservatives are apparently not of one mind in their assessments, and the skeptics within the religious establishment are joined by secular moderates who question whether the benefits of obtaining nuclear weapons are worth the political and economic costs of confronting the international community.

Augmenting these arguments is a potential security one. It is not a great leap from the logic that technology acquisition is a right and a national virtue to the assumption that technology is a national protector. In this regard, Hymans (2006: 2) speaks of "fear and pride" as two sides of the same coin—in his words, an "explosive psychological cocktail" that drives some leaders to acquire nuclear weapons. These influences—grounded together in a profound belief about the "true" nature of the nation and its relation to outsiders—complement and reinforce each other: fear motivates the seeking of nuclear weapons; pride suggests that the technology is within reach (Hymans 2006: 13, 35).

This is not to say that "fears" and "threat" are ungrounded in the "objective" universe. A security justification is that much more salient and sellable domestically in the presence of potential threat. It must be asked, then, whether the military concerns of Iran can justify its acquiring a deterrent nuclear capability. Iran repeatedly points to Israel's nuclear weapons as a threat to Iranian security (and the double standard by which Israel is permitted such weapons while Iran is precluded from acquiring them) but Israel and Iran have avoided *direct* conflict with each other. Another neighbor, Pakistan, also possesses nuclear weapons but relations between Iran and Pakistan are not adversarial (the Iranian nuclear program was aided, in fact, by the illicit purchase of equipment from Pakistan, albeit perhaps not its government). Iraq has been Iran's most formidable foe within the region, and the Iraqi threat of WMD (given prior Iraqi chemical attacks on Iranian forces) was not merely hypothetical for Iran in years in which the country's nuclear industry was under development. Still, the Iraqi threat has been effectively neutralized by the demise of the Hussein regime and Iran is positioned by merit of its Shiite religious character to influence the direction of the Shiite majority in the remaking of Iraq. Put simply, then, "Iran has no historic enemies; existential threats; or giant, hostile neighbors requiring it to compensate for a military imbalance with a nuclear program" (Chubin and Litwak 2003: 105). But this does not mean that the Iranian regime is blind to the unforeseen threat; indeed, the Iranian leadership consistently cites Hussein's attack on Iran in the early 1980s, and the phenomenal costs of that war, in arguing that Iran must prepare for the unexpected. Given US efforts to isolate the Iranian regime and US promotion of a policy of preemption, the Iranian leadership could well view nuclear weapons as a guarantor of regime longevity in the face of outside hostility and real threats to the survival of the Iranian regime. No matter how much the Iranian leadership feared and despised Saddam Hussein, they could not ignore an important lesson from the Afghanistan and second Gulf wars—that the US could quickly overtake a conventional military and drive a regime from power.

Unlike Iran, North Korea is involved in an intense rivalry and was the target, at various points in time, of implicit (and perhaps explicit) nuclear threats. The US used the nuclear threat against North Korea to counterbalance its conventional capabilities in suggesting that these weapons might be used in 1953 if Korean armistice negotiations failed and then when introducing these weapons into South Korea in 1958 to bolster its defense (Bleiker 2003: 725). Although these weapons were withdrawn in 1991, the North Korean regime still has reason to view its nuclear weapons as a counterbalance to US threats to employ nuclear weapons. The 2002 *Nuclear Posture Review* envisioned the possible use of nuclear weapons against North Korean targets to counter a conventional offensive or destroy WMD (Bleiker 2003: 727). Even without a US nuclear option, the Kim Jong-il regime appears to view nuclear weapons as a means for ensuring regime survival within a hostile environment. From the Korean standpoint, an invulnerable nuclear retaliatory capability—or at least a capability that stands a reasonable chance of surviving a US preemptive assault—presents a grave danger to South Korea and (directly or indirectly) the US, should the US try to eradicate the North Korean regime. Rhetorically, at least, North Korea has promoted this deterrence reasoning. In 2005, a North Korean spokesperson assailed the US for demanding that North Korea "give up [its] nuclear *deterrent* facilities [emphasis added]" without appropriate US concessions. The North Korean response was, "Don't even dream about it" (Kahn 2005).

Nuclear weapons also serve the interests of the North by providing a critical bargaining chip. With the lessening of tension on the Korean peninsula and the progressive isolation and impoverishment of North Korea, nuclear weapons provide leverage that prevents the North Korean regime from suffering death through benign neglect. It is North Korea's (likely) possession of nuclear weapons that gives North Korea front-page news status, grabs the attention of academic authors (such as myself) who are not regional specialists, and demands notice from a US presidential administration that is otherwise preoccupied with the war on terror. North Korea must fear the consequences should it eventually renounce its nuclear program. Lacking generous support from outside allies, threatened by growing cooperation between China and South Korea, and facing severe impoverishment and the likelihood of future economic crises, North Korea is obviously reluctant to place itself in a position in which "good will" alone binds the US and its allies to their promises.

Bargaining

The behavior of Iran and North Korea does not suggest that either country seeks nuclear weapons in an irrational drive to destroy their adversaries. In fact, in some respects, the behavior of the Iranian and North Korean regimes is reminiscent of the behavior of Iraq under the Saddam Hussein regime: all three countries have employed coercive tactics and have resisted inspections, which creates the impression that they have something to hide. Through their behavior, however, these countries reveal that they consider the benefits and costs of their actions. This is apparent from the conduct of both countries before and during the period in which the Bush administration was formulating its arguments about the necessity for preemptive war options.

North Korea has resorted to dramatic threats, implicit and otherwise, in its nuclear dealings with the US and its allies. This was apparent, in late 2002, when North Korea announced, after unsuccessful talks, that it was restarting its 5-MW nuclear reactor and resuming construction of two others, it expelled inspectors from the International Atomic Energy Agency (IAEA) and proceeded to remove surveillance equipment and seals on the reactor, spent-fuel rods in storage ponds, and reprocessing plants; it implied that it was to begin work on reprocessing the 8,000 spent fuel rods into plutonium; and it threatened publicly to withdraw from the NPT (Samore 2003: 16). The message was arguably more direct in the reported 2003 political aside to US Assistant Secretary of State Kelly[11]—and when North Korea claimed, on various occasions, that it possessed nuclear weapons, had reprocessed its spent fuel rods to produce plutonium, and/or was enriching uranium. That these are calculated confrontations is indicated by North Korea's hostile escalations yet reluctance to provide conclusive proof of its activities and retracting of announcements and softening of tone when this would serve the country's goals. In 1993, North Korea announced its intent to withdraw from the NPT rather than accede to UN inspections (under the treaty) that could disclose whether North Korea had diverted plutonium for nuclear weapons. It then suspended its withdrawal when the Clinton administration took an accommodative stance. A decade later, after apparently confessing a uranium enrichment program to Kelly, North Korea claimed that, in actuality, it had only a "right" or a "plan to produce nuclear weapons" (NTI 2006b). As Bleiker (2003: 729) observes, "it is striking how closely North Korea's approach in 2002–3 paralleled its behaviour during the crisis of 1993–4."

Thus, the evidence is substantial, as Cha (2002) argues so well, that North Korea has pursued a "coercive bargaining strategy" that aims to create a more favorable bargaining environment through the calculated and controlled use of threats and violence. Although coercion seems an unlikely choice for a weak country, coercion is a sensible strategy when pursued within limits and the weaker party seeks only to improve its bargaining position. The irony, then, is that weak states, facing a status quo that promises limited gains and significant losses, might initiate conflict with powerful states. As Cha (2002: 65) puts it:

> like the gambler down on his luck, the more North Korea frames the current situation as a losing one, the more appealing the double-or-nothing option becomes. Such a calculation would be based not on winning but on avoiding further loss.

This strategy can succeed for weaker countries in part because the stakes are higher for them than for their stronger negotiating partners, who must spread their attention and capabilities to address a variety of pressing global issues (on relative weakness and conflict, see Lebovic 2003; Lebow and Stein 1994: 361; Wagner 2000). Whereas the US is currently focused on its wars on terrorism and in Iraq, North Korea and Iran view the future of their nuclear programs as preeminent foreign policy issues.[12] A consequence of this asymmetry is that these countries have the stronger incentive to hold their ground. A coercive bargaining strategy can also succeed for these weaker states because they cannot easily be disarmed by force (as discussed below), and because they can reap great rewards from seemingly small concessions made by stronger states. These weaker parties know much more about their own nuclear programs than their bargaining partners do and can exploit insider knowledge to obtain concessions with hidden returns (e.g. preventing the disclosure of illicit activities by limiting when and where inspectors can operate). In addition, weaker states can mobilize international and domestic support through appeals to general principles that include "equality" and "fairness." Because the US is insisting that North Korea dismantle its nuclear program while European countries are conceding Iran's right to develop nuclear energy under international safeguards, North Korea is positioned to argue, as it has, that any US recognition of North Korean sovereignty that denies North Korea its basic right to pursue a "peaceful" nuclear program is a fundamental threat to North Korean sovereignty (Kessler 2005). To emphasize that very point, North Korea added a light-water reactor to its list of negotiation demands as "a matter of principle" (Cody 2005a).

Viewed from this bargaining perspective, then, it is unsurprising that progress in the six-party negotiations that commenced in 2003 under Chinese sponsorship was glacially slow—in the words of the US envoy, "excruciating" (McDonald 2005). The talks had been interrupted by a thirteen-month boycott by North Korea; and the third round of negotiations appeared deadlocked in the summer of 2005, well short of the stated goal of developing "agreed principles" that would govern further negotiations (Cody 2005b). The US took a hard-line stance in the negotiations reflecting its understandable reluctance to make commitments to a regime that violated agreements in the past—including the NPT, the 1992 denuclearization agreement with South Korea, and the 1994 Agreed Framework with the US—indeed, confessed to building nuclear weapons. At the same time, North Korea insisted that the US had not held to its prior commitments (for instance, continuing construction of the light-water reactors as promised in the Agreed Framework); demanded that it be allowed to retain its nuclear energy program; and asserted that it would only renounce its nuclear weapons program if the US made fundamental concessions up front that included reversing its "hostile policies" toward the North and a normalization of relations (McDonald 2005).

The point here is not that North Korea is an honorable negotiating partner and will abide by its commitments, nor that North Korea will concede its nuclear program through negotiations; there are reasons to suspect otherwise given the nature of the North Korean regime and the profound suspicion that separates the US and North Korea. The point is that North Korea has indicated, by its behavior, that it pursues a rational bargaining strategy, based on sensitivity to benefits and costs: its government has escalated conflict when it has suited its purposes and then adjusted its position when it obtained necessary and/or likely US political and economic concessions and the costs of failing to forge an agreement were high. Over the years, then, North Korea gave some ground when the US softened its rhetoric, indicated a willingness to respect North Korea "sovereignty," and held the promise of substantial economic rewards for North Korean compliance with US demands. Indeed, in September 2005, when all other options were seemingly exhausted, no further concessions were to be had, and China was pressing for compromise, North Korea accepted a statement of principles in which it promised to give up its nuclear weapons and programs, allow international inspection of its nuclear facilities, and rejoin the NPT (Kessler and Cody 2005). It did so knowing in all probability that it could not get more from the US than a vague concession that the US and North Korea would "discuss at an appropriate time the subject of" providing a light-water nuclear reactor to North Korea and that a failure to yield some ground would weaken the country's international political position.[13]

Iran has also not surrendered its goals despite extraordinary opposition from abroad. Its negotiations with France, Britain, and Germany acting for the European Union were unsuccessful given Iran's pursuit of nuclear autonomy and resistance to what Iran deemed "intrusive and illegal inspections" (Linzer 2005b). The Europeans insisted, for instance, that Iran import its nuclear fuel and export spent fuel; halt work on its heavy-water reactor, which can produce spent fuel for reprocessing into weapons-grade plutonium; ratify the Additional Protocol of the NPT, and commit not to withdraw from the NPT which permits international inspections of suspect sites (Associated Press 2005a). Although Iran has so far balked at these terms, even its turn to a confrontational approach does not reflect a fundamental abnormality in Iranian behavior. At the very least, Iran's attempts to retain its nuclear prerogatives and to avoid sanctions—conflicting at times and cooperating when necessary to prevent a firm and united opposition—establish that Iran, in pursuing its policies, is sensitive to costs and benefits.

Iran escalated conflict and acted defiantly to express dissatisfaction with outside demands. In August 2005, for instance, Iran resurrected a threat from earlier in the year and announced that it was removing the seals on its uranium ore conversion plant, which had been placed there by the IAEA. Shortly thereafter, Iran acted on its threat. Even then, Iran at least *tried* to allay concerns about its conduct. Iran announced that it would continue to freeze work on its uranium enrichment program (which could enrich the uranium hexaflouride gas to produce fuel for power plants or with additional processing, could produce nuclear weapons-grade materials), that it would store the gas under IAEA inspection, and that it intended to trade the gas for yellowcake with third countries (Fathi 2005).

Iran's behavior became more belligerent under the conservative presidency of Mahmoud Ahmadinejad. This is surely indicated by Iranian threats to withdraw from the NPT and the President's warning before the UN General Assembly in September 2005 that, with US provocations, Iran must "reconsider [its] entire approach to the nuclear issue" (Linzer 2005c). Nevertheless, Iran's behavior still did not reach the North Korean standard of repeatedly instigating major confrontations. This is true though Iran, in January 2006, removed the IAEA seals on a nuclear facility (at Natanz) to continue its "research" on uranium enrichment. The pattern over the years had been for Iran to pronounce its rights, to threaten resumptions in nuclear activities, and to resume some activities but to seek some level of accommodation with the IAEA and to retreat or equivocate to deflect efforts by the US and its European allies to refer the Iranian program to the UN Security Council for action (Linzer and Vick 2006). In private, at least, Iran continued to reassure foreign diplomats that reprocessing at that particular facility had not resumed and that Iran was still open to proposals that included reprocessing uranium in Russia (Moore 2006a). Public positions soon hardened: Iran insisted that its resumption of nuclear activities was "non-negotiable" and threatened retaliation in international oil markets should the IAEA refer Iran to the UN Security Council for possible sanctions; and the IAEA board voted overwhelmingly in February 2006 to do just that. But even as Iran hindered the inspections of its nuclear sites, announced that it would no longer be bound by the Additional Protocol of the NPT, and made inspectors remove their surveillance equipment from these facilities, Iranian officials maintained that Iran would adhere to the NPT, that Iranian research on uranium enrichment would not extend beyond the current program (Finn 2006a), in the absence of UN sanctions, and that the Russian reprocessing proposal remained an option (Dareini 2006).[14] Iran later offered to slow the country's nuclear program and schedule cooperation on some issues (Linzer and Moore 2006), and pledged to permit inspections should the Security Council hand the matter back to the IAEA (Reuters 2006). When the major powers agreed in June 2006 to provide a comprehensive incentive package to Iran to induce it to suspend its nuclear program until it could be judged peaceful, Iran refused to halt its uranium enrichment and threatened retaliation, once again, if sanctions were imposed on the country. Yet Iranian officials still spoke of the positive aspects of the proposal, in insisting for example that it required months of serious study before Iran could respond (Lynch 2006a), offered later to engage in far-reaching negotiations, and focused publicly on Iran's right to enrich uranium, not the problematic issue of intrusive inspections.

A broad look at the behavior of both Iran and North Korea provides further evidence that they are open to punishment. Iran has not behaved impetuously in dealings with adversaries. Instead, it has limited its physical attacks on Israel to support for (third-party) terrorism and accepted freezes and outside inspections that slowed the progress of the country's nuclear program. It has also foregone opportunities to punish its enemies. Iran could have attacked US vessels in the Persian Gulf, formed groups to engage in an all-out global war of terror against US and Israeli targets, and perhaps even adopted nonconventional (e.g. chemical) weapons for these purposes. This restraint is hard to reconcile with the assumption that Iran is anxious to employ WMD or to threaten to use them—that is, the assumption, more generally, that Iran is immune to punishment. The same can be said of North Korea, which has responded with signs of "pain" when the US has taken actions that have economically hurt North Korea (or at least some North Korean officials). For example, North Korea made the cessation of US efforts to combat North Korean money-laundering and counterfeiting a precondition for returning to the six-nation negotiations on the North Korean nuclear program (Kessler 2006a).[15] North Korea has betrayed its sensitivity to costs, as well, by acting with restraint against its neighbors. Although positioned to inflict catastrophic levels of damage on South Korea, North Korea countenanced the division of the Korean peninsula for over half-a-century. For that matter, North Korea has not used the nuclear weapons that the country might possess against South Korea or Japan—two potentially vulnerable targets.

Both North Korea and Iran most certainly recognize that the costs to these countries will increase should they use nuclear weapons against their adversaries. These costs obviously include the possible destruction of North Korean and Iranian society and their leadership in a (US or Israeli) retaliatory attack but also include costs inflicted by North Korea and Iran upon the intended *targets* of the attack. Efforts by North Korea to reunify the Korean peninsula through a nuclear offensive will amount (in the North Korean "best case scenario" of no US response) to a pyrrhic victory if South Korean cities and industries are laid to ruin and North Korea must live with the aftermath of nuclear war in territories that come under North Korean control, without aid from abroad (and a robust South Korean economy to provide support to the North). An annihilative attack by Iran on Israel, if limited to its centers of power, could still inflict mass casualties upon Israel's Palestinian population and risk destroying Islamic holy sites in Jerusalem (see Quester 2005).

The upshot, then, is that even secretive and conflictive behavior is insufficient evidence that Iran and North Korea are immune to the cost-benefit analyses that are the basis of deterrence. Neither country has exhibited the wild, uncontrollable actions that are expected of countries motivated purely by malice and revenge. Even the Bush administration has implied that these countries can ultimately be deterred. It suggested that coercion, if applied firmly and consistently, might *compel* Iran and North Korea to concede their nuclear options. Thus, it implied that these countries might surrender their nefarious commitments—publicly and irreversibly—given the *possible* political, economic, or military consequences of not relenting to outside demands. As significantly, the US shifted its attention away from North Korea in late 2005 and early 2006 when North Korea had seemingly acquired nuclear weapons and was developing (or extending) its capability to deliver those weapons against the continental United States. US concerns about the North Korean program subsided, then—not when the program became inherently less threatening—but when the US had to accept the inevitability of a nuclear-armed North Korea and to learn to live with the consequences.

Denial: "damage limitation" through preemption

This section examines the intelligence and operational challenges that hinder the effectiveness of a preemptive or preventative attack that is intended to disarm a fledgling nuclear power. These attacks became central to the US approach to the rogue-state problem with the September 11 attacks; and, by 2002, the idea that the US could not wait for its enemies to strike was enshrined in US military doctrine through a shift toward "preemption" in a number of high-profile US policy documents.[16] The principle was articulated in President Bush's State of the Union address and then the *National Security Strategy of the United States* (Bush 2002a), the *Nuclear Posture Review* (which among other things called for the US to develop low-yield, nuclear bunker-buster bombs to destroy enemy WMD that are buried underground; see US DOD 2002), and the *National Strategy to Combat Weapons of Mass Destruction* (Bush 2002b). (On these documents, see Kitfield 2005.) The principle was codified a few years later by the Joint Chiefs of Staff (US DOD 2005) in a draft doctrine (*Doctrine for Joint Nuclear Operations*) that gave US military commanders the option of requesting presidential approval to use nuclear weapons against those who employ or intend to employ WMD against the US, its allies, multinational forces, or civilians (Pincus 2005a, 2005b).

The challenges of preemption are considerable, however, both in the gathering of intelligence when preparing to attack and the effective executing of an attack. The challenges are that much greater when chemical and biological weapons are targeted. Compared to nuclear weapons, these do not spring from an obvious infrastructure and are insinuated more easily into civilian industry and scientific research (this is especially true of biological weapons). Attacking chemical and biological weapons stocks also requires knowledge of how the specific agents are stored if they are to be destroyed without spreading their deadly effects (Glaser and Fetter 2005: 95). That this section focuses on challenges of intelligence and destruction related to nuclear programs should be taken, then, as understating the problems should the US or other countries resort to preemption.

Intelligence problems and the limits of detection

The effectiveness of preemption ultimately hinges on the ability of the initiating country to pick its targets for maximum destructive effect. To quote David Kay, "the Achilles' heel of a doctrine of preemptive war or bombing strikes is that it requires really sound and complete intelligence, because if you can't precisely locate a target, you can't kill it" (Kitfield 2005). Good intelligence is likely scarce. As Henry Kissinger (2006) observes, "preemptive strategy is based on assumptions that cannot be proved when they are made. When the scope for action is greatest, knowledge is at a minimum. When knowledge has been acquired, the scope for preemption has often disappeared." Indeed, the attacker is likely to be planning a strike because the target country was less than forthcoming or purposefully deceptive when releasing weapons program information.

Intelligence collection and analysis is a tough business. This is clear from the many times that the US was surprised by international developments, such as the nuclear tests by India and Pakistan in 1998 when these countries declared themselves unambiguously to be nuclear weapons states and, in the same year, when North Korea demonstrated its three-stage missile capability by test-firing a Taepo-dong 1 ballistic missile over Japan. The challenges of detecting and locating facilities that could produce WMD, even nuclear weapons, are especially acute. A country can disperse its centrifuge-based, uranium-enrichment program and house it in relatively small facilities that require relatively little electricity (that could signal illicit activities). Or else, a country can bury nuclear facilities under ground, which would make it difficult to pinpoint the location of centrifuge equipment or processed nuclear material. As a proficient tunnel-builder, North Korea can easily hide nuclear weapons materials within the country's vast tunnel network. The US cannot see underground with reconnaissance satellites and unmanned aerial vehicles and it cannot obtain critical information about a suspect program from electronic intercepts. Thus, it might have to rely upon circumstantial evidence, such as the absence or presence of visible electric power lines (Richelson 2006: 351). Then, an adversary can manipulate the expectations of a foreign intelligence service, for instance, by employing camouflage, altering behavior around weapons sites, spreading false or misleading information, or procuring materials in ways that conceal their end uses. The reality is that countries are positioned to change the beliefs of US intelligence agencies or to reinforce their false assumptions because these countries might have a working knowledge of US intelligence sources and methods (derived, in part, from US disclosures of information about suspect activities). For example, India apparently used its information about US satellite orbits to time misleading behavior around weapon sites to when satellites were passing overhead, and incriminating behavior for periods when those sites were not being observed (Ellis and Kiefer 2004: 91–3).

Even apparently incriminating information can only disclose so much about a program. For example, evidence that centrifuges were purchased abroad will not reveal whether the equipment will be used in a nuclear program, much less betray the size and location of nuclear facilities. Similarly, heat outputs detected from thermal imaging satellites can point to the existence of an above-ground, gaseous-diffusion uranium enrichment facility but cannot reveal reliably whether it is producing low-enrichment fuel for a power plant or highly enriched material for a bomb (Bernstein 2001; Levi and O'Hanlon 2005: 63). In the end, intelligence agencies must extrapolate from what is known. For example, judgments about how much and when plutonium can be acquired from a nuclear reactor can derive from evidence of the reactor's operating capacity, which determines how much plutonium is produced and how long spent fuel must decay before it can be handled safely (Pinkston and Diamond 2005). In consequence, questions must remain about exactly what countries such as Iran and North Korea are doing and where they are doing it[17]—whatever the available evidence. Claims by the Bush administration (based on thousands of pages of alleged Iranian documents) showing that Iran had experimented with nuclear warhead designs still did not satisfy skeptics. They wondered whether the information was planted to deceive US intelligence or whether tests betrayed an actual effort to develop nuclear weapons. Likewise, evidence that Iran had planned a second uranium-conversion facility could be interpreted to indicate that Iran was intensifying its conversion efforts or, instead, that Iran was assessing the advantages of alternative conversion sites or a backup facility in the event the first one was bombed (Broad and Sanger 2005a; Linzer 2006).

The irony is that, for all the faults of US intelligence on Iraq, the US had considerable access to Iraq's programs via international inspection teams throughout much of the 1990s, compared to the access that the US now has in Iran and, certainly, North Korea. But it is now clear that the US, and others, were completely wrong in their assumption that Saddam Hussein possessed WMD. The assumption was expressed legendarily when the Director of the CIA, George Tenet, insisted that the case that Saddam Hussein had WMD was a "slam dunk." That belief permeated the US intelligence community. For instance, a 2002 National Intelligence Estimate expressed the consensus view within the community that Hussein had biological and chemical weapons and was reconstituting his nuclear program. As a result of the intelligence failure on Iraq, intelligence assessments of Iran and North Korea have become more qualified (Jehl and Schmitt 2005). The severe inhibitions to intelligence gathering and analysis demonstrated with respect to Iraq suggest, however, that it will be difficult to improve these assessments. These inhibitions were detailed in a report to the President from the US Commission on the Intelligence Capabilities of the United States Regarding Weapons of Mass Destruction (2005) that included a lengthy analysis of the pre-war failings of US intelligence on Iraq.[18] Its basic conclusions deserve repeating for they convey the immense intelligence challenges underlying preemptive war strategies.

First, the failures of intelligence on Iraq reveal the power and the danger of a false assumption. US intelligence agencies were inclined toward overestimating Iraqi capabilities given that the US had underestimated the extent of the Iraqi weapons program in the lead-up to the first Gulf War. By the time of the war, Iraq was apparently within six months to two years of developing a nuclear device and had ample stocks of chemical and biological agents (Ellis and Kiefer 2004: 9). Having been wrong, the burden of proof shifted now onto those who would deny that Hussein possessed WMD. The assumption that Hussein *did have* these weapons shaped all intelligence judgments. Thus, the absence of information was treated like confirmatory evidence; and ambiguous or contradictory evidence was interpreted to support the prevailing view. For instance, mixed evidence about whether Iraq intended to use imported tubes in an illegal centrifuge program or, instead, in a legal rocket program, was attributed to Iraq's efforts to promote a cover story. The result was that the prevailing argument was "nonfalsifiable"—it could withstand any and all evidence. Moreover, the prevailing view was not subject to rigorous testing through the gathering and assessment of necessary data. Analysts concluded that the suspect tubes were inappropriate for Iraqi rockets without knowledge of those rocket specifications, which were available and would have supported the opposite conclusion. In fact, the prevailing view benefitted, at times, from a "layering" that magnified the impact of weak evidence that supported the prevailing view. Conclusions were piled upon prior ones without regard for the quantity and quality of the information upon which the earlier inferences were based. According to a report by the US Senate Select Committee on Intelligence (2004), a judgment that a specific tanker truck was related to chemical or biological weapons activity was used to support a latter judgment that Iraq was involved in chemical weapons transshipment that was used, in turn, to support an assessment that Iraq had renewed production and stockpiled a huge amount of chemical weapons. As the report concluded (p. 29), the intelligence community "drew on very few pieces of new evidence to reach large conclusions in which new pieces of evidence would accrete to the previous conclusion and pieces that did not fit tended to be thrown aside." The consequence of these various dynamics was that analysts elevated plausible (hypothetical) uses of Iraqi technologies over likely ones and worst possible applications of these technologies over less threatening ones.

Second, pre-war failings reveal deficiencies in intelligence due both to an absence of human sources and faulty information from these sources. The US had no spies in Iraq after 1998 and its human intelligence drew largely from the activities of international inspectors that depend upon active cooperation from the target country. But even human sources are fallible, of course; and they can have political agendas that affect the veracity of information disclosed. To an astounding degree, US intelligence relied upon the (false) claims of a single source (aka "Curveball") linked to Iraqi exile groups. US intelligence agencies did an inadequate job of vetting their sources and failed to authenticate critical documents (such as the forgeries that implicated Iraq in an effort to buy uranium from Niger).

Third, the intelligence failings reveal that problems of analysis are compounded within an organizational context. Organizations create problems of compartmentalization. Due to the rigidities of organizational structure, important information, analytical resources, and viewpoints are not shared within the intelligence community. One result is that incomplete or flawed analyses are passed on to policymakers.[19] Organizations also make it hard to challenge existing inferences. As analyses are passed among agencies and up the chain of command, conclusions become "separated" from the supportive evidence (such as information about the number and reliability of intelligence sources). Consequently, those who use the data are not informed when "false" information is later recalled, hold to interpretations that are predicated upon inaccurate data, are prone to confuse "repetitious" with "corroborative" information, or misread the available data. Perversely, increased US intelligence gathering by satellite in the prelude to the second Iraq war convinced analysts that it was *Iraqi* activity that was increasing. The assumptions behind conclusions are also obscured. For instance, it appears that different organizations within government— e.g. the Department of Energy and the CIA—were reaching the same conclusions by relying upon incompatible sets of assumptions. Moreover, nuance, qualifications, and disagreements are lost in summary judgments. Indeed, dissenting views are missing from analyses or are relegated to "footnotes" so that they are unlikely to reemerge as the majority view. Within government, then, it becomes increasingly difficult to raise old questions or explore first principles once an intelligence judgment is rendered.

Even these problems minimize the challenges to devising sound policies because they assume that policymakers are neutral on the facts. The Commission found no evidence that members of the Bush administration tried to intimidate members of the intelligence community into producing analysis to support the case for a US invasion of Iraq. Although the US intelligence community was accused in prior administrations of enforcing a "party line" view,[20] there does not appear to be direct evidence of overt interference here. Still, there is considerable evidence that the administration exaggerated the case against Iraq by promoting the most dire view of the available data, emphasizing evidence that supported the case, and presenting Iraq as an imminent threat to the US. This could not have been clearer than in (then) National Security Advisor Condoleezza Rice's warning against being overly cautious when judging the facts: "we don't want the smoking gun to be a mushroom cloud." In addition, there is evidence that the administration influenced intelligence by creating a politicized atmosphere. A top intelligence official (Pillar 2006) in the administration asserts that analysts felt subtle pressure to give policymakers what they wanted, downplayed contrary information, held disconfirming evidence to a high standard, and conceded to the wishes of policymakers simply by allocating resources to the issues that were high on the administration's agenda.[21] At the very least, it can be said that the Bush administration— because it liked what it was receiving—did not actively challenge the intelligence community to defend its product. As the saying goes, "you don't ask questions if you don't want to hear the answers."

Thus, politics, organizational, and psychological dynamics conspired to undermine the intelligence information with which US policy was based, when the strength of preemptive policies hinges on the usefulness of this information. The challenges do not end, however, with data collection.

Preemptive options

Countries seeking to counter proliferation might have few good alternatives in an offensive-denial strategy that relies upon military force. Such a strategy invites onerous policy trade-offs. US policymakers try to underplay these trade-offs by referring to preemptive attacks when they mean preventative ones: it is easier to justify risky and costly policies when US adversaries presumably have both the intent and capability to attack the US or its interests with prohibited weapons. Still, it is hard to ignore the limitations of an offensive-denial strategy that is based in either preemptive or preventative force. This can be appreciated by assessing the relative merits and weaknesses of surgical strikes and internal intervention (regime change and/or territorial occupation) as counter-proliferation options.

Surgical strikes

A desirable military option for the initiator is the knock-out blow— disarming an opponent in a quick attack. The specter of such strikes has loomed over the current negotiating impasses with North Korea and, especially, Iran. This was clear when US Vice President Richard Cheney issued the not-so-thinly veiled threat that Israel might attack Iranian nuclear facilities if Iran were not more forthcoming in negotiations with Europe (a threat accompanied by widespread reports that Israel was preparing a preemptive attack if the negotiations fail, e.g. Mahnaimi 2005); it was also clear some weeks later when President Bush stated unequivocally that "all options are on the table" should Iran not halt its nuclear program. In case the subtlety was lost on the audience, he pointedly reminded Iran of lessons from the "recent past" in which the US had secured its security interests through force (Associated Press 2005b).

The effectiveness of any such attack depends in part on whether the blow is meant to be preemptive or preventative. Preemptive action allows the initiator to focus its attack on the immediate threat—weapons stored or positioned for use. Not only can the preemptor use its military forces efficiently, then, it can make a political case that it had to strike first to avert imminent disaster. In turn, preventative action offers benefits that include: (a) surprise and greater preparedness given the broad time band in which the initiator can act; (b) a variety of attack options (e.g. hitting one or more vulnerable links in the nuclear fuel cycle); (c) a visible and relatively concentrated target set (e.g. nuclear reactors and reprocessing facilities); (d) an ability to strike with a reduced fear that surviving weapons will be launched in retaliation; and (e) early intervention that provides time for historical developments to move in the initiator's favor. Unfortunately, for the initiator, these are hypothetical benefits and, in practice, the problems of preemptive and preventative measures are similar, and severe.

First, political complications will arise well before weapons arrive on their targets if overflight permission must be granted by countries bordering the targeted states. For example, Turkey, Saudi Arabia, and Jordan would not want to be seen as parties to an unprovoked attack on Iran (especially from Israel); and the US granting of overflight over Iraq for an Israeli operation against Iran, or the US use of bases in Iraq, would complicate the US effort in that country. Doubts about the success of any operation will only make it more difficult to gain the cooperation of neighboring countries. India is claimed to have denied a request from Israel that its planes be granted landing and refueling rights for a proposed attack on a Pakistani nuclear reactor partly out of fear that the reactor would not be destroyed (Moriarty 2004: 75). Enduring the political and military fallout from an "unprovoked" attack is hard enough, enduring fallout from a failed effort is intolerable.

Second, attack plans will be devised in all likelihood around problematic and deficient intelligence. Targeting requires knowledge of the multiple locations in which illicit materials and suspect weapons are located, with no allowance for error if weapons that were not destroyed could be launched in retaliation against the attacker or a more proximate target (Israel or South Korea). The poor US track record in locating and destroying both fixed and mobile Iraqi Scud missile launchers from the air in the 1991 Desert Storm operation is not heartening. Mobile launchers were easily concealed and could quickly fire their weapons and move to new locations to avoid counterattacks. Thus, despite hundreds of sorties flown to destroy these missiles in the first Gulf War, at least half of Iraq's fixed launchers appear to have survived the war and not one US hit on evasive mobile launchers was confirmed. Indeed, lacking an exact count of these launchers, the US was unable to gauge its success in destroying them apart from the reduced number of missiles that were being fired by Iraq over the course of the war (Ellis and Kiefer 2004: 171–4). Nor is the US ability to target suspect biological, chemical, and nuclear sites during the first Gulf War a reason for confidence. Although the US estimated that the large number of strikes against Iraqi WMD targets had severely degraded Iraq's nonconventional capabilities, the results of UN inspections later revealed that a substantial portion of Iraqi sites had not been targeted—in fact, only a small fraction of nuclear targets had been struck and most biological weapons facilities had been spared (Ellis and Kiefer 2004: 170–1). It seems that targeting remained a challenge in the second Gulf War, despite years of scrutinizing Iraqi facilities in the interlude between the two wars. The post-war occupation revealed that many facilities were wrongly identified or no longer active (Glaser and Fetter 2005: 91). US adversaries have only become better at deception in light of their own lessons learned from these conflicts.

Third, the success of an operation will be hindered by the protection of targets. The nuclear infrastructure in Iran and North Korea is well-dispersed, buried at depths that could impede the effectiveness of even deep-penetrating weapons, and designed around redundant (and perhaps hidden) facilities and equipment. Weapons and production capabilities are also protected when "hidden in plain sight" and interspersed with civilian storage and manufacturing facilities (Glaser and Fetter 2005: 93). Thus, the destruction of centrifuges, reactors, and, reprocessing facilities might still not disable the Iranian nor North Korean nuclear programs. At best, an attack on these facilities would *delay* Iran's acquisition of nuclear weapons (see Kam 2004). In the case of North Korea, the likelihood is that the country has built secret facilitites, for instance, that could produce weapons-grade uranium. These facilities can be small in scale and protected underground or in caves that could well be immune to attack with earth-penetrating conventional weapons. Even if North Korea's reactors and reprocessing facilities could be destroyed, the US could not destroy whatever weapons North Korea had stored or its enrichment program (Saunders 2003).

Fourth, the success of an operation will be mitigated by collateral damage. Optimistically, nuclear facilities could be destroyed in aerial strikes and their deadly contents contained within collapsed building structures (Levi 2003), but the release of harmful nuclear, chemical, or biological materials into the surrounding environment is an ever-present danger. Destroying buried facilities could also pose a high risk to co-located populations. Conventional, earth-penetrating (bunker-buster) weapons cannot drive deep enough into the earth nor generate sufficient blast to strike many buried targets. Nuclear-armed penetrating weapons permit the destruction of more-deeply buried targets but not without endangering the surrounding populous through exposure to radioactivity, nuclear fallout, blast, and heat. Although penetration reduces the size of the explosive device needed to destroy a target, penetration cannot occur to depths that will contain the effects of a nuclear explosion. Moreover, the yield of nuclear penetrating weapons must remain substantial, given limits to a weapon's penetrating ability, and must grow non-linearly with the depth that targets are buried. Hundreds of thousands of people, if not a million or more, could be killed in a surgical strike on a country's buried facilities depending on the location of the target relative to populated areas. Indeed, even a nuclear explosion might not incinerate biological weapons in underground storage; the resulting blast could send the biological materials into the atmosphere extending the death toll (Medalia 2005; National Research Council 2005).[22]

Fifth, the initiator of the attack must contend with political fallout from a preemptive or preventative attack when viewed by other countries as a brazen violation of national sovereignty. The political consequences will be worse still if foreign personnel (e.g. Russians working in Iranian nuclear facilities) were among the casualties. The political fallout from the attack could well hurt efforts to obtain global cooperation to obtain access to suspect facilities under the NPT, to implement effective embargos of countries suspected of building illicit programs, and, indeed, to build support for the principles of non-proliferation upon which preemption is based. An "unprovoked" attack would only make it easier for the targeted state—or other states that are considering the acquisition of nuclear weapons—to make the case that those weapons are needed for self-defense. The political fallout would increase astronomically if the initiating state employed nuclear weapons for enhanced effect against deeply buried arsenals and/or facilities.

Sixth, the initiator must be concerned about the effects of precedent setting. Any such attack risks eroding normative constraints on the use of preemptive force. If other states do not share the belief that the targeted country posed an imminent security threat to the attacker (satisfying the international legal condition of "necessity") or that the size or nature of the attack was appropriate given the threat (satisfying the condition of "proportionality"), the effect of the attack could be to legitimize acts of preemption apart from considerations of self-defense as defined by the traditional international legal standard (see Arend 2003). The danger is that other countries will be that much more inclined, as a result of the precedent, to solve their "security" problems by acting subjectively, precipitously, and without restraint in the name of self-defense. Precedent setting would be a more likely outcome if the initiator attacked with nuclear weapons. Given the simplicity and clarity of existing prohibitions on the use of nuclear weapons (Kier and Mercer 1996: 94), such an attack would have the frightening consequence of "demystifying" nuclear weapons and breaking a taboo on their use that has been reinforced through over half-a-century of nuclear abstinence. The implications of violating the long-standing taboo should not be underestimated. Without appreciating the stigma associated with the use of nuclear weapons in the post-World War II era, it is hard to explain why relatively few countries have actually pursued the nuclear option and why, even in wartime, nuclear-armed countries have not used their nuclear weapons against non-nuclear adversaries. Cold War-era war-fighters might have regarded nuclear weapons as quantitatively *but not qualitatively* different from other types of weapons; the behavior of national leaders suggests that they accept the distinction (see Quester 2005; Tannenwald 2005).[23]

Seventh, the initiator must contend with the military consequences of an attack. A strike against nuclear facilities in Iran or North Korea will not disarm these countries. North Korea has built a formidable conventional army in hardened positions that is poised to take the offensive against South Korea. Indeed, the South Korean capital, which lies a short distance from the demilitarized zone, is an inviting target for North Korean artillery and missiles; and North Korea has threatened to respond with "total war" to any such attack (Samore 2003: 18). Iran's military is less formidable but has the capability to fire missiles at Israel and US bases in Iraq, to disrupt oil traffic in the Persian Gulf, and to draw upon a significant comparative advantage—support for terror groups throughout the Middle East and militants within Iraq's Shiite population. That the US has not been a principal target for these groups could change.

Eighth, any such attack is likely to exacerbate the security dilemma of the attacking state by creating a "self-fulfilling prophesy" based in a destabilizing fear of surprise attack. This problem was familiar to Cold War-era theorists (see Lebow 1987). If leaders believed that an attack on their capabilities was inevitable, they might seek to reduce their losses and obtain possible benefits by being the first to strike. Certainly, there are good reasons (discussed in the prior section) to assume that these leaders would be reticent to go on the offensive but there are also reasons to assume that leaders that go on the *defensive* will lose control in a conflict. For instance, leaders could disperse weapons and pre-delegate their launch authority to subordinates, which could then increase the probability that these weapons will be fired by accident or without due cause. (These possibilities are discussed in Chapter 4.)

Finally, any successful attack could fail in the long term. Israel's 1981 attack on Iraq's Osirak nuclear reactor in a preventative strike—though condemned widely at the time—is now generally regarded as the model for surgical operations—in its apparent quickness and effectiveness. A generous interpretation of the lessons of that attack is that it can work with a rare confluence of circumstances, that is, "specific and highly accurate intelligence and a negligible risk of collateral damage and retaliation" (Chubin and Litwak 2003: 109). A less-generous interpretation is that the operation was ultimately unsuccessful. Instead of destroying the Iraqi program and ending Iraq's nuclear ambitions, the attack apparently had the opposite effect. The evidence is that the Israeli attack convinced Hussein that his nuclear program was necessary, and his program was sent into "overdrive" (Salama and Ruster 2004). With an enormous increase in the budget and personnel, Iraq adopted a more efficient, less risky procurement strategy: it dropped the reactor for producing plutonium and moved directly toward enriching uranium in dispersed, redundant, fortified, and hidden facilities (Moriarty 2004; Salama and Ruster 2004).

Internal intervention

Another means for stripping a country of its non-conventional military capabilities is internal intervention. This can be accomplished by establishing some degree of political control, territorial occupation, or both, in the targeted country. Given the recent intervention in Iraq, the political and territorial options appear to be interlinked but that need not be true. Intervention can follow the highly intrusive model of the post-World War II occupation of Japan and Germany, and now Iraq, in which the initiator attempts to remake a country's political system, rebuild its economy, and/or establish civil society. Intervention can also take the less intrusive forms of an assassination of a hostile foreign leader, efforts to oust an unfriendly regime, or a short and limited occupation of a country or its suspected weapons sites.

Bringing a quick end to a political problem through an assassination or forced change of government is a desirable strategy from the standpoint of the initiator if it requires but a small expenditure of resources with an opportunity for large gains. Indeed, it is the ease with which an assassination can be ordered that has made it a source of domestic US and international consternation: the fear has been that assassinations will be used liberally, injudiciously, and reciprocally (e.g. a plot by Saddam Hussein to kill George H.W. Bush could have originated in prior efforts by Bush to kill Hussein).[24] Targeting leaders seems less politically (and ethically) problematic when the method of attack (aerial bombing) preserves the pretense that leaders were military or incidental targets, that is, "they were in the wrong place at the wrong time." But assassinations can backfire (a matter discussed again in Chapter 5); and they invite problems similar to those of other forms of political intervention: they cannot work well without a viable replacement government.

Seen from the US standpoint, the complication here is a Catch-22 of sorts. On the one hand, an opposition strong enough to take power in the first place might be too strong for the US to control, and can plot a course apart from US interests. Cases in point are the Afghani warlords within the Northern Alliance. They abetted the US effort to unseat the Taliban yet were able to deflect pressure to relinquish their ammunition stocks and fiefdoms, once a new central government was in place in Kabul, and to escape accountability for crimes and abuses they committed in the period prior to Taliban rule. On the other hand, an opposition that is weak enough to require US support might be too weak to assume power or retain political control. For example, Iraqi exile groups were too factionalized, marginalized, and powerless to present a meaningful threat to Hussein's governance without US intervention. They had neither the military capabilities nor popular following to force their way into power. The use of exiles to topple a government appears an even less inviting option against the entrenched political leadership of Iran and North Korea.

If an externally supported opposition did manage to come to power, it might always be tainted by its connection to the intervening force. In consequence, the new leaders might try to disassociate themselves from their foreign patron. It is instructive, in this regard, to recount the US relationship with Ahmad Chalabi who helped build the case for US intervention in Iraq. Given his apparent embrace of US interests and perspectives, Chalabi was a candidate favored by the US Department of Defense (and others within the Bush administration) for leadership in a post-war Iraq. Once in a position of leadership, he ran afoul of US decision-makers because of his apparent illicit business dealings and revelations that he had supplied phony intelligence to the US to justify US intervention. As troubling were Chalabi's efforts to distance the Iraq interim governing council from the US and double-dealings through which Chalabi was alleged to have passed sensitive US intelligence information to Iran. Whether the new leaders are weak or strong, then, the result might be disappointing for the intervening country.

Given the challenges of regime change, an intervening country could seek to limit its intervention to territorial control of suspect sites or other areas of importance to illicit programs (e.g. airports or seaports)—content to leave the offending government in power. The operative logic here is that to ensure the destruction of hidden weapons stockpiles and/or the infrastructure for research, development, and production of these weapons, it is necessary to occupy and search suspect storage sites and research, development, and production facilities. Assuming that occupation is not a health risk (from ambient radiation, chemicals, or biological materials), such an approach positions the intervening country to investigate suspect sites (by searching computer disks, viewing written documents, examining laboratories and equipment, and interrogating personnel), control access to those sites, use information to identify new sites, and eventually to destroy an entire weapons infrastructure.

At least, this is how it works in principle. A limited occupation could be quite problematic because the intervening country loses its military advantage after the initial attack. As occupying forces take their positions, the host government can rally public opposition, counterattack the (perhaps vulnerable) positions of the occupying army, and lash out abroad by employing force against neighboring countries or engaging in terror attacks on the intervening country or its allies. The host will also be on alert for subsequent occupations of suspect sites (after all, the host knows where they are). Intervening forces could have great difficulty moving among suspect facilities and might arrive there only after equipment or weapons have been removed or sites have been militarily fortified. Thus, instead of offering a reasonable (middle) alternative to the risks of a surgical strike and outright national occupation, a limited occupation might offer the "worst of both worlds"—an enduring involvement in a country paired with a limited ability to locate, pass among, and destroy critical targets. In the end, the dearth of intelligence, dispersion of targets, and military options available to the targeted country, which doom a surgical strike, could undermine a "surgical occupation."

An alternative then is some variant of the "Iraqi model"—the outright military and political occupation of a country suspected of building WMD. The difficulty here certainly varies with the nature of the operation—how long the intervening country plans to occupy territory, whether it intends to remake the institutions of the host country, its ability and willingness to rely upon local support in the effort, and the extent to which the occupier and public within the host country accept the principles behind the occupation— whether the host country sees the intervention as a "military occupation" or an exercise in "nation-building," and whether these terms have positive or negative connotations for the host. In fact, the difficulty of the task will likely increase exponentially with the ambitiousness of the intervening country's goals.

In the short-term, any attempt to occupy countries such as Iran or North Korea must be planned carefully as a full-scale operation with knowledge that the intervention: (a) will be anticipated by the target country; (b) will lead to acrimonious international debate; (c) will require a huge commitment of military resources by the intervening country; (d) could produce high fatality levels and great destruction within the target country and neighboring states (should the target country go on the offensive, e.g. North against South Korea); (e) can lead to a prolonged insurgency in which government forces or their opponents press their claims to power; and (f) could provoke the target country to use its surviving WMD. In fact, the inducement to use these weapons is never greater than when in the hands of a regime that is about to fall to an invading army. The regime's motivation to use these weapons is high (the regime is attempting to fore-stall its imminent collapse), the costs are low (the regime no longer fears that using illicit weapons will provoke intervention), and the invading army effectively provides a human shield behind which the regime can hide: then, it would be fratricidal for the invader to activate the "assured destruc-tion" option in retaliation for a nonconventional weapons attack.

The problems for the intervening country do not end with short-term military success. At a minimum, the intervening country must cope with a possible humanitarian crisis (i.e. death, misery, dislocation, and disorder) arising from the military intervention or power vacuum created by the top-pling of the defeated government. The speed with which the immediate crisis passes will depend upon the availability of food, clean water, and san-itation; medical supplies, facilities, and personnel; and security from looting and criminal, political, and retributive violence. It will depend further on the country's long-term capacity to recover and develop—the quality of local transportation, reach and capacities of indigenous government insti-tutions, the local availability of electric power, construction materials, financing for reconstruction and development, and technical expertise. The intervening country might invest in both short- and long-term solutions. Apart from any obligation the intervening country feels it has to address the humanitarian crisis, it must certainly fear that a failure to address deeper economic and political problems and grievances within a society will produce new security threats that will require intervention again at some point in the future. Indeed, any intervention is likely to be undertaken with sufficient seriousness that it is paired with some bigger ("nobler") purpose such as bringing democracy or political stability to a troubled country and area of the world. Even a mission that is initially limited in scope can grow into a serious entanglement. The occupying country will find that it cannot address short-term problems without investing greatly in long-term solu-tions—building roads, constructing and/or reconstructing sanitation facili-ties and electric power generators, and so forth—nor address long-term problems without investing greatly in short-term solutions—such as pro-viding security to stem the looting of new and old equipment and attacks on workers, managers, contractors, and government officials. A balancing of short-term measures and long-term solutions is one of many difficult and critical trade-offs that the intervening country is required to make.[25]

Despite the dangers of analogical thinking—mistakenly treating all future interventions as like a salient case (Khong 1992)—the disappointing experience in Iraq is instructive. It is useful if only because the lessons learned in Iraq build on the UN nation-building experiences of the 1990s—in Cambodia, Somalia, Haiti, and elsewhere—that show that the task of political and economic nation-building is formidable (Ottaway 2003; Paris 1997). The reality of such interventions is that they must succeed when a central government (as in Somalia) is lacking; when profound class and political conflicts exist within society (as in Haiti); when unsavory characters and groups (e.g. Afghani warlords and Iraqi Baath party members) vie for power and must be appeased to prevent their defection; when parts of the country remain in a state of war (in Iraq and Afghanistan); when neighboring countries (e.g. Iran) will attempt to exploit political instability to advantage; when corruption is endemic to a country and its government practices; when government officials and personnel most equipped to handle local problems are tainted by their association with the prior (e.g. Baathist) regime; when the "good guys" and the "bad guys" look alike to intervening government officials who lack local language skills, cultural sensitivities, and country expertise; when the public sector is bloated, corrupt, and mismanaged and the private sector is too undeveloped to carry the service load; and when economic remedies, social transformations, political measures and reforms, or government restructuring can exacerbate profound tensions within society. Tensions can increase no matter how limited or far-reaching these actions appear in scope. A seemingly innocuous or at least fairminded decision by coalition authorities in Iraq to reallocate public access to electricity, fueled resentment from those who had profited disproportionately under the Hussein regime without producing offsetting support from those who benefitted from the redistribution (Henderson 2005: 5). "Bigger" decisions come with greater risks. Constitutional debates in Afghanistan and Iraq must address divisive issues such as the position of women within society, the role of Islam within government, and how power will be divided among indigenous groups, which could bring new threats to national stability. For instance, a constitution designed around a federal structure risks the further alienation of Iraq's Sunni population, sandwiched between the oil-rich (Kurdish) northern and (Shiite) southern provinces; conversely, a strong central government in Baghdad risks exacerbating ethnic/religious conflicts among Sunni Muslims, Shiite Muslims, and Kurds, which could lead to political deadlock, at best.

Even this pessimistic portrayal of the challenges to occupation assumes a population that at least momentarily welcomes the intervening force as "liberators." It is a perverse footnote to the US operation in Somalia that the initial US reception in that country prompted the US military to speak of favorable lessons learned that could be applied to future US operations (Richburg 1993). There is little evidence that the populations (let alone military units) in Iran or North Korea would welcome US forces: a need to resist "foreign interference" and to oppose a foreign occupation is the one thing around which a divided public can unite. Then, things will be worse for the occupier that goes into an operation "feet first," without appreciating the challenges of intervention from the start. The US record in Iraq—and unlearned lessons from prior US occupations[26]—do not bode well, for the US, in this regard. The US initiated the Iraqi occupation, itself, based on faulty intelligence. As a result, the US failed to anticipate the threat posed by a possible insurgency, the huge amounts of weapons stashed around the country, and the dilapidated state of the Iraqi infrastructure (Gordon 2004).

Whether or not the occupier chooses to avoid the long-term entanglement of a commitment to nation-building, the historical record does not promise a successful outcome. By one account, only seven out of twenty-four military occupations were successes in the 1815–2003 period and six of these successes came with the end of the Second World War (Edelstein 2004: 57). Given the many common features of these contemporaneous cases—a perceived threat from the Soviet Union, exhaustion in war, eradication of the political opposition, dedication to reconstruction and rebuilding, and so on—it is impossible to draw definitive conclusions from them except, perhaps, that these cases *were* exceptional. In fact, what could be pluses for the occupier in some contexts could be negatives in others. For example, it can be argued that cultivating local leadership makes an occupation more tolerable for a local population. Yet this assumes a local leadership that is competent, willing, and able to work with the occupying force, not one that seeks to promotes its own interest or to undercut the occupier from any acquired positions of power (as is becoming increasingly true in Iraq). It also assumes that the local population will not take limited control as a signal that the occupier has a soft and/or finite commitment to the occupation (Edelstein 2004: 67–8). Among other complications, this could inspire opposition forces to ride out the occupation rather than seek common ground with local adversaries within emerging governmental institutions.

In short, it is unlikely that the US will find quick and easy solutions in the *deux ex machina* of regime change or surgical occupations, much less military occupations that could leave the intervening country without allies to help shoulder the burden. Ironically, too, these consuming efforts might be no more successful in addressing the immediate threat posed by WMD: a perversity of the US operation in Iraq is the limited attention that invading US forces initially paid to securing suspect weapons sites from looters, terrorists, and combatants. In assessing available counter strategies, it is useful to reflect upon the prescient warnings of a prominent deterrence theorist (Jervis 2003: 318) just before the second Gulf War: "a world in which allowing Saddam to build his WMD would be very dangerous is not necessarily one in which overthrowing him would be relatively cheap." There is every reason to suspect that this admonition applies even more so to Iran and North Korea.

Punishment and denial: economic sanctions

Given the problems of preemptive strikes and internal intervention, countries can employ economic sanctions to prevent others from acquiring WMD. Available sanctions include curtailing aid to the offending country or freezing its assets abroad. These methods can be effective. For example, the US used aid leverage, in the 1970s, to halt South Korea's nuclear program and, in the 1990s, to halt Israel's expansion of settlements in the occupied territories (Drezner 1999: 2, 254–75). The success of these specific methods could hinge, however, upon preexisting cooperation or a high degree of interdependence between the sanctioning and target countries. Consequently, much of the debate over the wisdom of sanctions centers instead on the utility of trade blockades and embargos. Blockades are intended to interdict goods and resources in transport to or from the target country (in the manner in which US ships boarded Iraqi-bound vessels and the US coastguard intercepts drugs intended for distribution within the US). Embargos operate through the domestic institutions of participating countries: countries use legislative and police powers within their own borders to stem the flow of goods and resources to and from the target country.

Both approaches can be employed unilaterally or multilaterally and can vary in extensiveness.[27] In other words, a blockade or embargo can target specific items that are related to an illicit weapons program, valued by the target government or its elite supporters, vital to the offending country's economy, and/or vulnerable to disruption; or they can target a large variety of goods and non-commercial exchanges (e.g. tourism). In addition, either approach can be employed to *deny* a country a war-making capability or to coerce (*punish*) a country into making concessions. When used coercively, the operative assumption is that concessions will follow indirectly when citizens hold their government responsible for the societal costs that sanctions inflict or directly when government leaders or their supporters suffer the effects of sanctions professionally or personally, for example, via travel or import restrictions. But coercion is effective, too, if it just convinces a government that it is likely to encounter unacceptable risks and high costs in pursuing an illicit program. Therefore, *denying* countries economic or military resources through blockades and/or embargos can serve a punishment strategy that is intended to *deter* a country from pursuing a weapons program or to *compel* a country to concede an existing one. It makes sense, then, to address sanctions simultaneously as both instruments of denial and punishment.

Sanctions are controversial—first and foremost, because critics claim that they do not work. That case is strengthened, to an extent, by evidence from pre-war Iraq. The revelations of abuses, bribes, and profiteering involving businesses, governments, and UN officials under the 1996 UN oil-for-food program (which allowed Iraq to sell some oil to pay for food, medicine, and war reparations) is reason enough to doubt the effectiveness of efforts to restrict trade with a target country. Even the degree of success achieved by sanctions against Iraq (discussed below) in obtaining its compliance with some UN demands is attributable to the "exceptional" nature of the Iraqi case. Iraq was dependent on oil exports that flowed largely by sea: blanket prohibitions made it easier to determine whether trade in certain goods was prohibited under the sanctions (for example, Iraq was not allowed to pursue a peaceful nuclear energy program), participation in the sanctions regime was extensive, and substantial international attention and resources were devoted to containing the development of Iraq's weapons programs. Whereas the Iraqi case involved "the most comprehensive US and UN trade and financial sanctions regime mounted since the Second World War" (Hufbauer *et al.* 2001), its lesson is arguably not what sanctions achieved but rather how much was required to produce less-than-complete success.

Economic sanctions have allegedly had a poor record, overall, in bringing offenders into compliance with international standards of behavior. The success rate for sanction cases even when the US was part of a larger coalition of sanctioning states was around twenty-five percent from the 1970s through the 1990s (Hufbauer *et al.* 2001);[28] and there are reasons to suppose that sanctions will fare no better in halting weapons proliferation. Although Libya was under US sanctions from 1986 and then partial UN sanctions from 1992 (for its support for terrorism), it was still able to acquire sophisticated centrifuges, enriched uranium, and a design for a nuclear weapon from a secret network linked to Abdul Q. Khan, the father of Pakistan's nuclear program. The poor success rate of sanctions is attributable to a number of factors.

First, effective embargos and blockades are hindered when potential participants have strong reasons for impeding the imposition of sanctions or circumventing them. South Korea, China, and Japan (the latter, at least prior to North Korea's 2006 missile tests) preferred cooperating with North Korea and resisted the US confrontational approach toward that country. South Korean leaders even voiced support for a peaceful North Korean nuclear energy program. In turn, Russia and China are linked economically to Iran and are loath to confront it. Russia has aided Iran's nuclear reactor, missile, and air-defense programs, and China, with its large and growing energy needs, has major energy importation deals with Iran. Thus, both countries fought US efforts within the IAEA to refer the Iranian case to the UN Security Council (Baker 2006) and, within the Security Council, restricted their support, then, to a non-binding resolution that called upon Iran to suspend enrichment activities and cooperate with the IAEA—expressly written to avoid the suggestion that Iranian noncompliance would trigger sanctions or a military response (Lynch 2006b). These countries were joined on the IAEA governing board and UN Security Council by developing countries that were sympathetic to claims that Iran's rights were being abridged by rules governing nuclear technology acquisition that did not apply equally to all countries—termed "nuclear apartheid" by many, including Iran's new president in an address to the UN (Anderson 2005).

The US, itself, has undercut its own sanctions to serve various US national security policy objectives. A desire to obtain influence in South Asia and balance against China led the US to revise its stance toward Indian proliferation. The US offered to rewrite existing US laws and provide sensitive nuclear technology to India despite its acquisition of nuclear weapons, its refusal to sign the NPT or subject all of its nuclear reactors to international inspection, its apparent intent to upgrade its nuclear weapons capabilities, and its vocal support for Iran's civilian nuclear energy program (Linzer 2005d; Kessler 2006c).

Likewise, a contribution to the war on terror motivated the US to lift sanctions against Pakistan for its nuclear activities, and later allowed Pakistan to escape US criticism for its non-cooperation in investigating (and the country's possible role) in the A.Q. Khan network. The US went easy on Pakistan, despite Pakistan refusing to allow UN inspectors to question Khan, who supposedly spread nuclear technology to countries such as Libya, Iran, and North Korea, took "business trips" to a dozen-and-a-half countries, and disseminated detailed plans for fabricating nuclear weapons, including manufacturing secrets that would take competent countries years to uncover on their own (Broad and Sanger 2004, 2005b). The US even sought to circumvent its own non-proliferation sanctions to serve non-security policy objectives. The US Congress moved to amend the 2000 Iran Non-proliferation Act that prohibited US purchases of Russian space technology (in retaliation for Russia's exports of nuclear technology and missiles to Iran) when the US was forced, with the grounding of the US space shuttle, to rely upon Russian space craft to access the international space station (Gugliotta 2005).

Second, targeted countries can learn to do without scarce goods; and some countries have exhibited a high tolerance for international isolation and economic deprivation. This seemed true of Iraq, which, under sanctions, suffered a huge decline in gross domestic product, exports, and imports (O'Sullivan 2003: 122–36). It is certainly true of North Korea, which has held to its positions though teetering on the verge of economic collapse. Indeed, sanctions arguably hurt the wrong people within a targeted country. It is said that government leaders and elites find ways around sanctions—even get rich and powerful off them (Andreas 2005)—allowing contraband to flow into a country; and that the poor, disenfranchised, and helpless (e.g. small children) suffer disproportionately when sanctions bring a loss in tourism and trade, or otherwise damage the economy. In these countries, the suffering public might be badly positioned to push for policy change. Not only do countries such as North Korea lack a democratic opposition that can bring pressure upon the government, their leaders can undermine the opposition by using sanctions to justify a state of emergency, economic deprivation, and the suppression of dissent (Andreas 2005: 341).

Third, the effectiveness of embargos and blockades is compromised when participating countries make it hard, if not impossible, for the target country to comply with terms that will end sanctions. For sanctions to work, countries must be punished for their nonparticipation, and *rewarded* for their participation in a weapons control regime. Such rewards, which can take the form of a lessening of punishment, can be slow in coming. In the case of Libya, the US was understandably reluctant to allow the free flow of resources to and from a country that had previously supported terrorist movements and, by any reasonable standard, is hardly a democracy. For a host of reasons, then, including suspicions about Libya's weapons programs, the US kept its sanctions toward Libya in place, and abstained in the vote on the 2003 UN Security Council resolution that lifted sanctions, after Libya made amends for its role in the bombing in the late 1980s of two international airliners including a US Pan American flight over Lockerbie, Scotland. There was actually considerable resistance within the US government to changing its stance toward Libya (Jentleson and Whytock 2005/06: 71–4). US objectives could also have impeded the performance of sanctions against Iraq: because an implicit US goal in Iraq after the first Gulf War was regime change, Iraqi compliance with UN resolutions calling for transparency and disarmament would fail to satisfy the US. Hussein might have gained little from agreeing to the terms of the resolutions.[29]

Fourth, the effectiveness of these sanctions hinges on the target country's actual need to acquire resources abroad. Substitutes could be available, at home, for items that are in scarce supply; or else countries can do without global purchases entirely. The extent to which North Korea developed its nuclear infrastructure with outside help is a matter of dispute; what is not in dispute is that North Korea has reached the point of no return in the nuclear fuel cycle. It can no longer be denied a nuclear weapons capability through the deprivation of essential program materials.

Fifth, the effectiveness of sanctions is limited when banned products are easily sold or acquired on the global market. Products (e.g. oil) from the target country are sometimes in great demand and exports (e.g. centrifuges) to the country will flow when the market price compensates for the costs and risks of trade. The Khan network was based in a large number of countries that did not have or enforce laws restricting nuclear technology exports, and it relied upon companies that were more concerned with profit than how the product would ultimately be used. These governments and companies became part of the network specifically because they were "poorly prepared to resist the Khan network's lucrative offers" (Albright and Hinderstein 2005: 120). Even with formal limits on trade, goods can flow into and out of the target country. Exporters and importers can often rely upon a multitude of smuggling routes and methods of transportation, especially when prohibited goods are relatively small, lightweight, and easily transported. In addition, importers can employ subterfuge to obscure the end uses of procured items. Iran built its nuclear program by playing on weaknesses in export control regimes—by setting up front companies, buying items in limited numbers, and acquiring dual-use technologies (Howard 2004: 103–6).[30]

Sixth, the effectiveness of sanctions is limited when offending governments can conspire with one another. North Korea used its knowledge in an illicit technology area (i.e. missiles) in trade for knowledge in another (i.e. nuclear fuel reprocessing) with Pakistan. Although the Pakistani government claims that A.Q. Khan acted independently in his dealings with North Korea, the possible size and quantities of material that were shipped to North Korea make it unlikely that Pakistan's military was not involved. Pakistan's president has suggested that the transfers to North Korea included many centrifuges, centrifuge parts, and some amount of uranium hexaflouride (Masood and Rohde 2005).

Seventh, sanctions might be slow to affect the target country's economy. Under Security Council Resolution 661 in 1990, the UN imposed sanctions of unprecedented severity on Iraq to force its withdrawal from Kuwait. These sanctions banned all trade, oil imports, financial transactions, and arms sales and included a freezing of Iraqi assets and, later, were paired with an air and naval blockage (Cortright and Lopez 2000: 40–1). Even these severe actions could not bring changes in Iraqi policies within the acceptable time frame; military action was taken against Iraq less than six months after its occupation of Kuwait.

Finally, the effectiveness of embargos and blockades might be encumbered when those sanctions are most needed—in halting a fairly well-developed weapons program. Compared to deterrence, compellence is hard to achieve because it requires states to make visible concessions and to give up what they already possess (see Chapter 2). Both Iran and North Korea have balked at what they perceive to be outside efforts to limit their options (peaceful or otherwise) and rights as sovereign nations; and North Korea will surely resist dismantling its well-developed nuclear infrastructure given the costs and the long-term loss in leverage in dealings with outside countries that will result. Even acceding to UN inspections requires that these countries *give up* some level of security by releasing compromising information. It should not be forgotten that data obtained by UN inspection teams helped the US target suspect facilities during the second Gulf War.

Notwithstanding these problems, blockades and embargos have been used against illicit programs with some success. Iraq's attempts to acquire dual-use technologies were often thwarted by the US and its allies; and severe economic sanctions in effect against Iraq since its invasion of Kuwait in 1990 deprived Iraq of financial resources for pursuing WMD programs. There is evidence, for example, that Iraq's inability to acquire uranium after the UN Security Council passed Resolution 687 in 1991 contributed to the demise of Iraq's nuclear program (Salama and Ruster 2004). Moreover, Iraq did concede to many of the demands made in Resolution 661: Iraq partially or fully complied with seven of the resolution's eight conditions (Cortright and Lopez 2000: 55). For example, Iraq eliminated its nuclear weapons capability (and was believed by the IAEA to have done so). The fact is that Iraq *did* ultimately disarm; it just did not disarm with the transparency that the UN required. The evidence suggests, in fact, that the compellent effects of Iraqi sanctions were far-reaching. They "compelled Iraq to accept inspections and monitoring and won concessions from Baghdad on political issues such as the border dispute with Kuwait"; in addition, they "dramatically reduced the revenue available to Saddam, prevented the rebuilding of Iraqi defenses afer the Persian Gulf war, and blocked the import of vital materials and technologies for producing WMD" (Lopez and Cortright 2004). Even more convincing evidence of the effectiveness of sanctions is found in the Libyan case. In 2003, after years of outside economic and political pressure, Libya announced that it would end its WMD programs, come into compliance with major arms control treaties (including the NPT and Additional Protocol), and open the country to full inspection. Since then, Libya has allowed unprecedented access to its weapons facilities, technologies, documents, and stockpiles that include enriched uranium, ballistic missiles, and chemical weapons and has aided the dismantling and destruction of its nuclear, chemical, and missile programs. The magnitude of the Libyan turnabout was clear from the extensiveness of the Libyan nuclear infrastructure which included HEU, a reactor program, centrifuges of varying sophistication, and a design for a bomb (Squassoni and Feickert 2004).

Of course, the effectiveness of sanctions is again a matter of intense international debate as the US and its allies have sought to punish North Korea for its nuclear test and Iran to bring it into compliance with the NPT and Additional Protocol. Because Iran has not been isolated politically and economically, policymakers have good reasons to doubt whether the Libyan model can be applied to Iran with a favorable result. Yet, the Libyan case is instructive. In various respects, it shows that sanctions can succeed despite strong reasons to suspect they will fail.

First, UN sanctions against Libya were fairly limited in scope and, perhaps, in immediate impact. The sanctions focused selectively on Libyan tourism, air travel, diplomatic movement, and airline and oil industries. Given the interests of Libya's trading partners, the UN sanctions permitted Libya to market its oil—the major source of the country's export earnings.

Then, the sanctions were not universally respected and were increasingly violated. But sanctions deprived Libya of technology necessary for developing its airline and petroleum industries and created uncertainty about the long-term stability of the Libyan economy, which discouraged private investment (O'Sullivan 2003: 117, 200, 218). These economic effects complicated Libya's efforts to contend with rising unemployment, inflation, and popular discontent (Jentleson and Whytock 2005/6: 66).

Second, the sanctions took years to take effect. They were the result of patience, perseverance, and extensive lobbying by the US and its allies to obtain international support for applying increased pressure on Libya. For example, given the desire to obtain a unanimous verdict in the UN Security Council to increase the legitimacy of UN action against Libya, the US, UK, and France waited for the council terms of Cuba and Yemen to expire (Hurd 2005: 506). The sanctions regime tightened progressively through resolutions passed when Libya failed to comply with prior ones.

Third, just as Iran has bolstered its international political position by citing a fundamental right to nuclear technology, Libya achieved some success through appeals to international liberal principles of fairness and due process (Hurd 2005). Still, Libya was effectively isolated by UN sanctions, the building of support for them, and the worldwide response to the country's flagrant challenge to the UN Security Council. Although Libya had supported terrorism and pursued WMD to obtain a regional and global leadership position, instead it acquired international pariah status. As Hurd (2005: 504) notes, sanctions "had an enormous impact on the status and prestige of the government in its dealings with IOs and other governments." Libya was forced, in fact, to devote considerable energy to defending itself before the international community. The combined effect of sanctions on Libya was to raise significantly its political and economic costs of doing business in the world.

It is impossible to surmise from the Libyan case, or from other recent episodes, that sanctions can be fine-tuned to bring maximum political pressure on a government while sparing its citizenry, nor that such targeted or "smart" sanctions are necessarily desirable. Indiscriminate sanctions are more effective at imposing costs than smart sanctions that target the wrong goods or industries, produce "pinprick" effects, or cannot prevent target governments from reallocating resources to further their priorities, perhaps, at the expense of the public. For example, the narrow targeting of a country's weapons imports could cause the country to transfer resources from social-welfare to military programs to cover the increased costs of procuring illicit weaponry (Drezner 2003). But the Libyan case adds to the growing evidence that sanctions work best when they are imposed with modest goals, mechanisms for monitoring global adherence and target compliance, and a greater amount of international cooperation rather than less (O'Sullivan 2003: 284–320).

Sanctions are potentially more effective when paired with other policy tools. These include diplomatic initiatives in which the offender is confronted with a *threat* of sanctions (or additional sanctions) before the country learns to live without targeted goods and resources. Iran's efforts to obtain a political settlement with European negotiators was surely motivated in part by a sanctions threat. The effectiveness of sanctions is arguably multiplied, too, when joined with (implicit or explicit) threats to use military force to impose additional costs on the target country. These threats could be useful if only because they might induce uncommitted countries to participate in sanctions to prevent the eruption of violence. But military threats (and the actual use of force) interpreted as bluffs can damage the threatener's credibility and bargaining position (as discussed in Chapter 5); and threats taken seriously can provoke the target government and its citizenry, exacerbating a crisis and strengthening the government's domestic position and political resolve. These threats appear least efficacious when most likely to be used—against a government that is committed to developing a nuclear infrastructure specifically to produce weapons. If nuclear weapons are what a government wants and its worst-case scenario is that it cannot acquire these weapons, it has nothing to lose from holding out. Indeed, a destructive attack on the weapons program could offer dividends to the target government in the form of increased domestic and international support and maybe a loosening of sanctions, helping the target government realize its nuclear goals. By holding out, the government stands at least some chance of weathering the storm and presenting a fait accompli, in the form of a finished bomb, to the global community. The India, Pakistan, and (perhaps) North Korea cases suggest that, once a country has the bomb, others will accept it.

Importantly, too, sanctions can be paired with incentives for compliance (Newnham 2000)—offers of trade benefits, economic aid, and assistance to the target country for developing its peaceful nuclear enterprises. These incentives could be used to compensate the country for the (economic, military, and political) costs of dismantling a nuclear program or to encourage the country to comply more fully with agreements that require nuclear inspections and data sharing. These are important goals inasmuch as a paradox of the NPT is that it requires the active cooperation of participating countries to verify treaty compliance. Certainly, the Libyan experience points to the value of pairing sanctions and incentives: the promise of a better future as much as the continuing threat of isolation induced Libya to end its support for terror organizations and pursuit of nuclear weapons. Agreement was possible once Libya received concessions that included a trial of the Libyan terror suspects in a Scottish court in the Hague (rather than an international court), an end to efforts to pursue "regime change" in Libya, and a normalization of diplomatic relations. Here and elsewhere, sanctions appear most effective "when they are linked to incentives as part of a carrot-and-stick bargaining process" and are "understood as instruments of persuasion rather than punishment" (Cortright and Lopez 2002: 15–16). A reward strategy can help overcome a major challenge for a punishment strategy: rewards can facilitate compromise when countries are least inclined to cooperate. After all, the poor state of relations between the target country and its accusers is why sanctions are being considered in the first place.

It must be said, though, that the North Korean case offers inconclusive lessons, at best, on the utility of incentives: whereas an offer of rewards by the Clinton administration ended the impasse with North Korea in the early 1990s (see Drezner 1999: 275–304), North Korea ultimately reneged on the (1994) agreement. Still, without the agreement, the size of the North Korean nuclear force could have grown considerably in the years between 1994 and 2002. North Korea froze the reprocessing of thousands of spent fuel rods into bomb-grade plutonium and the construction of two large reactors that could have produced enough plutonium to build dozens of bombs a year. Indeed, the agreement might not have been abandoned in full by North Korea had the US rigidly adhered to the terms of the treaty and not accused North Korea of violating it (Newnham 2004).[31] Construction of the light-weight reactors that were promised to North Korea was years behind schedule, and US reluctance to normalize relations with North Korea and efforts (however justifiable) to pressure the North Korean government for its poor human rights record and involvement in international criminal activities placed North Korea on the defensive. At worst, the North Korean case suggests that a reward strategy can create conditions that aid a coercive strategy. The light-weight reactor program, like other benefits that were extended to North Korea, increased its reliance upon the US and its allies as sources of fuel, financing, spare parts, training, and aid (Howard 2004: 820–1).[32] These dependencies are a potential source of future leverage.

The success of embargos and blockades in controlling the spread of illicit weapons depends inevitably on how "success" is defined. Sanctions might "work" at some level though not inducing compliance with specific demands within a designated period. On this point, Baldwin (1985: 240–3) notes that sanctions could be effective when isolating the offender for purposes of international consensus-building (see also, Crawford and Klotz 1999), signaling still other countries not to follow the offender's example, and imposing costs on third parties that seek to aid the offender. Illustrating the possible harm for these parties is the 2003 interdiction by German and Italian authorities of the BBC China carrying centrifuges bound for Libya, which helped to unravel Khan's global nuclear supply network. More generally, Baldwin (1999/2000: 92–3) observes that sanctions do not simply succeed or fail. Not only must sanctions be judged more favorably "the greater the effectiveness, the lower the costs for the user, the higher the costs of noncompliance for the target, the higher the stakes, and the more difficult the undertaking," sanctions must be judged more or less favorably relative to available alternatives. Thus, in response to the suggestion that the success rate of sanctions has been low, the question to be asked is, "low compared to what?" All things considered, the record of sanctions—and accompanying policies of political isolation—compare favorably to the recent record of military intervention.

Conclusions

US policymakers overstate the military effectiveness of preemptive (offensive denial-based) strategies against the nonconventional capabilities of so-called rogue states. Even the destruction of these capabilities will not disarm these states and could leave them with troublesome retaliatory options. At a minimum, a preemptive attack could increase public backing for a weapons program and provoke the target government to intensify its acquisition efforts and shun international cooperation. To do otherwise is to risk appearing weak and losing domestic support.

By the same token, threats to use force *in retaliation* for an attack do not invite the costs of preemption. Although the very factors that confound efforts to disarm states in a first strike could limit the military effectiveness of a second strike, rogue states must believe that unlimited and sustained retaliation will be the natural consequence of any nonconventional weapons attack upon the US, or its allies. In consequence, these states must harbor some doubt that their small nonconventional arsenals (and links to them) can survive a retaliatory attack. They must also recognize that adversaries will be far less concerned about inflicting collateral damage and incurring costs of intervention when these countries believe their fundamental interests are threatened and that a wrong must be avenged. The US action in Afghanistan sends a powerful signal: after the attack on the US World Trade Center, there was no doubt that the US would settle for anything less than excising al-Qaeda from Afghanistan. The behavior of rogue states and the logic of their situations suggest that these states can read these signals. Whatever they might believe they can gain from an attack, they can hardly ignore the prohibitive risks of unprovoked aggression that could take the lives of hundreds of thousands of Americans.

Although preemption *promises* quick and effective results against countries that are isolated, strong-willed, or impervious to pain, denial strategies are poor substitutes for punishment strategies in addressing the rogue-state proliferation problem. If the attacker is unable to assess post-attack damage to suspect facilities, to determine the existence of duplicate facilities, and to know whether damaged technology and materials were repaired or replaced, the attacker must inevitably fall back on a coercion strategy to obtain information on the suspect program and impede its rehabilitation and growth. Denial tactics could actually hinder the implementation of a punishment strategy. After an attack, leaders of the target country might strongly resist sharing information that could be used for future attacks on the country's facilities. Their will to resist is likely to be stronger still if they sense that undermining the country's leadership was the actual purpose of the prior attack.[33] Such resistance is troublesome however effective was the preemptive attack. It is worth remembering that the exodus of UN inspectors from Iraq in the late 1990s created a short-fall of information that fostered worst-case thinking about Iraqi intentions and capabilities (Lopez and Cortright 2004). The result was a costly operation in Iraq from which the US has yet to extricate itself.

It must be added that the threat of reprisal might be the only means available to prevent the proliferation of weapons to countries that have historically abided by the terms of the NPT. Once a country can enrich uranium or produce plutonium, most of the "hard work" has been done; then, the country could withdraw from the NPT, halt inspections, and move to produce bombs (Einhorn 2004: 25). In consequence, the US and its allies could be left without denial-based options for preventing the country from acquiring (or using) these weapons.

Admittedly, deterrence might be ineffective in preventing countries from acquiring chemical and biological weapons, which are easily hidden, or even in preventing their use. In this vein, Koblentz (2003/4) asserts that biological weapons confer offensive advantages over the defense that could undermine deterrence: it is difficult to defend a population when the large number of available agents and delivery systems make it challenging for the defender to know what and how pathogens will be used in an attack. But deterrence has always functioned with the presumption that the offense is advantaged over the defense. It is for this reason that US policymakers relied upon deterrence over defense against nuclear weapons in the Cold War years. If anything, biological weapons, more so than nuclear weapons, bring uncertainties of use—effects that are "delayed, variable, and difficult to predict" (Koblentz 2003/4: 105)—that play to the assumptions of deterrence.[34] These effects include collateral damage should the effects of the biological attack spread back to the initiator and the possibility that the victimized country will retaliate massively once determining responsibility for the attack.[35]

To say, however, that rogue states can be deterred is not to say that the US should accept, as inevitable or desirable, the prospect that nuclear weapons (or other nonconventional weapons) will proliferate to additional parts of the globe. There is always a possibility that these weapons will fall into the hands of terrorists or safeguard the rule of unsavory regimes. Beyond this, the possession of nuclear weapons by any country presents an existential threat, in the sense that it increases the chance that, under some conditions, these weapons will be used. Rationality will not necessarily prevail in times of conflict and any government can succumb to the "use-them-or-lose-them" logic of preemption (a perhaps mistaken belief that a decisive advantage goes to the country that is first to strike), a belief that all has been lost that justifies retributive actions, or a false sense that the conflict is under control, which then leads the government to take unnecessary risks. Indeed, the disquieting flip-side of the North Korean coercive bargaining strategy is that North Korean leaders might use their weapons somehow to send strong signals that cause a conflict to spiral out of control. The North Korean test firing of a large number of missiles, including a long-range Taepodong-2 missile and nuclear test in 2006, despite world pressure to forego the tests, suggests the ease with which a major confrontation can ensue. (These possibilities are discussed in more detail in Chapter 4.) The fact is that rogue states will not always act in their best interest. After all, Hussein did miscalculate in ways that ultimately led to his downfall; and the North Korean leadership arguably resists domestic economic and political changes that could relieve internal and external pressures upon the regime. But the US and its allies retain options in combating the proliferation threat. These involve the use of diplomatic, economic, and/or military actions, in some combination, to deny states access to offending technologies or to punish states to obtain their compliance with prevailing arms control regimes. Whether or not countries will dismantle their weapons programs, these options promise deterrence benefits if only by slowing or constraining program progress or growth.

Policymakers must recognize, however, that the worst-case scenario for the US and its allies is not that a rogue state will come into possession of a nuclear weapon. Instead, the worst case is that, *because of* preemptive efforts, a war erupts for which the initiator is unprepared—perhaps a war in which the offending weapons are actually used. Military force is always an option against weapons proliferators—in fact, a good option under hypothetical conditions of imminent threat, a promise of success, and/or limits to the negative consequences. In the current confrontation between the US and its rogue-state adversaries, these conditions have not been met.

4

A DEFENSIVE-DENIAL STRATEGY AGAINST ROGUE STATES

National missile defense

In taking office in 2001, the Bush administration moved quickly to deliver on its campaign promise to build a missile defense system to protect the US from attack. For the administration and its supporters, such a defense was a rational (if not moral) response to fledgling nuclear adversaries—especially Iran, North Korea, and Iraq (before 2003)—that eschew traditional rules of restraint and possess small arsenals against which the US could credibly defend. Of course, the September events of that year forced a profound redirection in US policy away from planning against an adversary, with a fixed base of operations, that would adopt the "traditional" Cold War-style of attack—firing missiles over intercontinental distances against US targets—to a geographically dispersed adversary that could sneak into the US and deliver its blows through any number of creatively destructive means. But the case for missile defense is only strengthened, proponents argue, by the horrific terrorist attack on the World Trade Center in New York, which proves once and for all that the US faces a new nemesis, one that will go anywhere and do anything to accomplish its goals.

In making their case, proponents have muddied the relationship between long-standing US deterrence policies and missile defense. Some proponents insist that defense and retaliatory deterrence are complementary responses to the new threat but that defenses can "negate the potential of regional adversaries" and render an attack "not only fatal, but futile" (Slocombe 2000: 80).[1] Defense enthusiasts (including the President and congressional Republicans) make an even stronger argument. They maintain that defenses will eradicate the nuclear threat from leaders that cannot be deterred by the threat of punishment (retaliatory deterrence). Indeed, one commentator went as far as criticizing opponents of missile defense for "sporting Cold War arguments so comically anachronistic you have to wonder whether they'll *ever* offer a serious argument on nuclear weapons" (Krauthammer 2001). In contrast, for cautious supporters of a limited NMD system, defenses must be balanced against deterrence: defenses offer some insurance should deterrence fail, but defenses can undercut deterrence if hey overreach their potential or aggravate the global nuclear threat (Daalder

72

et al. 2000; Lindsay and O'Hanlon 2001).[2] It is useful, then, to rediscover "lost" Cold War deterrence arguments, to juxtapose them with current thinking about missile defense, and to explore the implications of these assertions given the likely performance and vulnerabilities of an NMD system. Through this analysis, the dilemmas and risks of NMD become apparent: to an extent unappreciated even by cautious supporters of a limited NMD system, defenses do not replace retaliatory deterrence, and can undermine it.

In the Cold War period, AD advocates typically voiced the view that defending US cities against a missile strike was unnecessary, impractical, and provocative. It was *unnecessary* because the deterrent potential of the US threat to retaliate against the Soviet Union was sufficient to prevent a Soviet attack. It was *impractical* because even the most optimistic assessments of the potential capabilities of US defensive technology would still allow thousands of Soviet warheads to rain destruction upon US cities. Indeed, the US would need to invest vast economic and technological resources in a missile defense system to produce but a small and arguably inconsequential enhancement in US security. Even that improvement would disappear quickly if the Soviets sought to compensate by building more missiles inasmuch as the rules of competition in nuclear armament favored the offense. In turn, a defense was *provocative* because it promoted a mistaken belief that defenses could significantly reduce a country's exposure to retaliation. If the Soviets believed that US defenses gave the US an advantage, the Soviets might arm to offset those defenses or strike the US before its defenses were in place.

In contrast, many AD critics (i.e. war-fighters) embraced efforts promising to protect US cities from a Soviet nuclear attack. Strikingly, though, even basic assumptions shared by these critics were revised to make the current case for missile defense. These assumptions include: (a) that the adversary can be deterred with the right offensive policies; (b) that the adversary would most likely tailor its attacks to restrict the opponent's retaliatory options; (c) that missile defenses, at best, complement nuclear deterrence; and (d) that extending deterrence to allies is an especially difficult task. Because these assumptions give rise to a variety of retaliatory deterrence arguments (a wider variety than AD advocates prefer),[3] they guide this chapter.

The chapter proceeds as follows. First, it assesses the conditions under which the US might face adversaries that cannot be deterred by US offensive options. Second, it assesses whether adversaries can and will use nuclear weapons to coerce the US given its inclination and ability to retaliate. Third, it presents a simple statistical analysis of the effectiveness of the defense against various offensive threats to determine, among other things, whether the residual threat with an effective defense might still require a US deterrence policy. Fourth, it compares the effects of attacks on less-defended (non-US) targets with those on well-defended (US) targets to gauge the incentive that adversaries have to attack the former if an effective US NMD system is in place. The chapter concludes that NMD advocates have not made their case that Cold War-era deterrence principles are ill-suited to US relations with fledgling nuclear states. Specifically, it concludes that the expected value of a missile defense of any type and scale is low because it is unlikely to enhance global security above levels offered by retaliatory deterrence.

The "conditional" undeterred adversary

Most NMD proponents share the apprehensions of pessimists who believe that the stability of US–Soviet relations cannot be replicated consistently when nuclear weapons proliferate to various parts of the globe (Karl 1996/7). There are certainly reasons (discussed below) to believe that a stable balance is unlikely to emerge between the US and its new global adversaries. What sets most NMD proponents apart, however, is their decided emphasis on the danger to stability posed by ruthless, irrational, or self-destructive leaders.

For many NMD proponents, the worst-case scenario of a ruthless and irrational leader armed with nuclear missiles and ready to launch them seems closer to being realized with the apparent recklessness of the September 11 attack on the World Trade Center. President Bush, for one, was motivated to ask, "Suppose the Taliban and the terrorists had been able to strike America or important allies with a ballistic missile?" (Graham and Allen 2001).

As troubling as that possibility is, it is not clear that government leaders with the mentality and policies of the Taliban or al-Qaeda could thrive in a country that possessed the technological resources and know-how to develop and shelter a nuclear missile capable of striking the US, nor even that these leaders could import and assemble the variety of components needed to construct such a weapon. It is also improbable that such leaders pursue their objectives without concern for benefits and cost. Their self-interested instincts and behavior are noteworthy because a land-based missile cannot be fired at the US without revealing the country from which it was launched and could well bring swift and devastating US retaliation. An attack by missile over intercontinental distances is a virtual invitation for retaliation, as it comes, by merit of ground and satellite tracking, with an easily and quickly recognized signature. Therefore, when NMD proponents ponder the implications had the Taliban possessed ballistic missiles, they should also ask, "Suppose the Taliban knew what it knows now—that an unprovoked, deadly attack on US territory would bring a US response that would soon force the Taliban into exile?" Would the Taliban then have allowed al-Qaeda to co-opt the instruments of the state and control its destiny? As previously suggested, there are strong reasons to answer in the negative.

The problem though is that in overstating the danger posed by unconventional leaders, NMD proponents understate the danger should *any leader* unwittingly lose control of its weapons.[4] Indeed, NMD proponents are on stronger ground when they shift their emphasis from the peculiar predispositions of rogue-state leaders to general circumstances that could lead states, under any leadership, toward non-rational behavior. The strategic question changes productively, then, from how to counter evil leaders to how to control conditions that can lead to dangerous behavior. It is the latter question that is addressed in the section to follow. It first discusses a possible rogue-missile attack arising from: (a) a fading opportunity to attack—a possibility that NMD proponents acknowledge, and (b) strategic conditions, which NMD proponents tend to ignore.

Fading opportunities

NMD proponents recognize that leaders facing imminent destruction might do the unthinkable—fire missiles at US targets even if provoking an annihilative US response. The logic is that leaders about to surrender power have nothing further to lose and no reason for restraint; thus, they might take high risks or even commit suicide, in grand style, by taking the enemy with them. This logic cannot be dismissed out of hand. After all, it echoes the logic of AD doctrine that the US would commit suicide by launching an all-out retaliatory attack on the Soviet Union in response to a Soviet attack. But this doomsday scenario is based on an unlikely and perhaps even illogical set of political conditions, for a number of reasons.

First, even AD proponents did not assume that the US would launch its missiles when *anticipating* an attack or when the result would fall short of the *destruction* of the adversary. Put simply, AD was a deterrence doctrine. Policymakers promised the *worst* hoping they would never have to deliver on the promise; they would trigger the doomsday sequence only in response to an *actual* adversary attack. Even then, AD proponents recognized that the decision to commit suicide by retaliating for an attack could be a tough one. It was for this reason that US policymakers tried to automate the act of US retaliation that would bring the Soviet death blow, for example, by pledging publicly and repeatedly to use these weapons if attacked.

Second, the logic behind the scenario does not reflect a profound understanding of a "new threat." Some hawkish AD skeptics voiced similar concerns about how the Soviets would behave under threat. They expected an attack on the US as the "last gasp" of a dying Soviet regime (Gray 1986: 85); and they were wrong. Despite the prominent view in the Reagan administration that Communists would cling tenaciously to leadership (see, e.g. Kirkpatrick 1982), the Soviets left power with a whimper, not a bang.

In hindsight, it seems that US analysts had overestimated the stake that Soviet leaders had in continued leadership, underestimated their willingness to live to fight another day, and overstated the importance of the US in the day-to-day political life of the Soviet Union.

Third, the logic whereby a nuclear strike on the US can help or satisfy a failing leadership begs for examination. It is not enough for NMD proponents to claim that rogue-state leaders will take abnormal risks because they "value regime survival above all" (Slocombe 2000: 80); the causal principles that link regime survival to a nuclear attack require some elaboration. Proponents must consider that political or institutional rivalry and hopes by individuals or factions to secure their place in a new government might counter reckless attempts by rogue-state leaders to take their countries into nuclear war; that leaders hostile to the US have strong reason to keep the US out of the fight, not invite its unrestrained wrath; and that the devastating consequences of US nuclear retaliation for a rogue-state nuclear strike would realize the worst nightmare of an unpopular regime—massive instability and a total breakdown in political control. In consequence, then, any decision to launch nuclear missiles might be postponed by rogue leaders who continue to control the instruments of domestic power and war-making (e.g. they can launch nuclear missiles). Indeed, the point at which it is rational for those leaders to launch their missiles is the point at which they have nothing left to lose and all their options have been exhausted—the point at which it might already be too late for action.

Fourth, the extreme scenarios that could lead rogue states to an irrational missile attack actually assume that those states are rational and conservative. This is true, for example, of the scenario (immortalized in the movie, *Dr Strangelove*) in which a regime ensures, somehow, that its weapons will be fired even with the collapse of the regime. This is a troublesome possibility that deserves (and, in the next section, receives) attention. On this point, one expert claims that, during the first Gulf War, Saddam Hussein ordered his military to attack Israel with nonconventionally armed missiles if communications with Baghdad were severed and it was about to fall (Baram 2001). A government report (Duelfer 2004) asserts further that Hussein ordered biological weapons attacks on targets, including Israeli cities, under conditions (presumably extreme) that have yet to be made clear. But the logic behind the automated response, in which the crossing of some line triggers a rogue-state response, is the logic of limited objectives. By drawing a line, a state signals that it values some things (e.g. survival) more than others and that it will take the ultimate step only when those values are threatened. After all, Hussein accepted a substantial conventional defeat in Kuwait without unleashing his chemical arsenal.

Fifth, even extreme scenarios do not mean that rogue states will inevitably attack, and the US retains political and military options to prevent a missile launch even in time of war. The US prepared for its final march into Baghdad anticipating that Hussein might order the use of nonconventional weapons in his possession. The result was intense US efforts to convince Iraqi commanders and their subordinates to ignore any launch commands: Iraqi personnel were warned that they would be held accountable individually for their actions and that they had little interest in tying their future to the soon-to-be-vanquished Hussein regime. In addition, the US adopted tactics that could drive a wedge between Hussein and his military commanders. US bombing waves against Baghdad were interspersed with periods of calm in which these military leaders could reconsider their position toward the Hussein regime.

In sum, NMD advocates are correct to suggest that conditions could lead states to do the otherwise nonrational; their thinking here is a commendable departure from brazen assertions that rogue states simply cannot be deterred. Even then, these proponents exaggerate the extent to which these conditions will induce an "irrational" response.

Strategic dilemmas

Fears of an undeterred adversary reflect a narrow conception of the threat. Whereas NMD advocates play up the threat from rogue leaders with malevolent intent, they discount the threat to deterrence should defenses move these leaders to increase or protect their military options by relaxing control over their weapons. If some combination of US offenses and defenses was believed by US adversaries to impair their first strike or retaliatory options, these countries could adopt offsetting actions that cause their weapons to be fired accidentally or prematurely. Thus, defenses could exacerbate a deterrence problem that results *despite the intentions* of adversary leaders. In strategic terms, accidental or unauthorized weapons use is a *deterrence stability* problem caused, in part, by the interdependence between offense and defense: just as US offenses can be used defensively to limit the retaliatory damage that an adversary can inflict, US defenses can be used to limit the retaliatory consequences of a US offensive, that is, a preemptive attack.[5]

Complementary US offenses and defenses present severe challenges for US adversaries with small arsenals and undeveloped weapons controls, and only increase the chances, then, that rogue states will use weapons by accident or without authorization. The danger exists if rogue states take a variety of compensatory actions. Consider three possibilities.

First, a rogue state armed with a small number of vulnerable missiles (e.g. launched from fixed sites) can attempt to improve the chances that these weapons will survive an attack and penetrate the defense by decentralizing launch authority—delegating it to military leaders or their subordinates. That way missiles can be fired even if communication links (e.g. with civilian leaders) are severed in combat.

Second, a rogue state can position weapons close to targets to circumvent US defenses (and protect perhaps against the destruction of these weapons by US offenses). For instance, these states can deploy missiles on ships in forward positions or else they can send agents, under a cloak of extreme secrecy, to pre-position a nuclear device on US territory for detonation, if necessary, at some point in the future. A pre-positioned device could be small (although perhaps not small enough to have earned the label, "suitcase bomb"; see Lindsay and O'Hanlon 2001: 76), highly lethal in a populated area, delivered with *relative* ease given the porousness of US borders and the density of the US population along the US coastline, and used without immediately revealing the identity of the attacker.

Third, a rogue state can ensure that the *largest* number of people are capable of employing a weapon but can design a weapon (and activation procedures) to permit its use by the *smallest* number of people in time of need. This makes sense from an efficiency standpoint. By initially widening the pool, the military can increase its flexibility to act: if some trained individuals are ultimately unavailable for a mission, a military can rely upon other personnel. By equipping weapons to be easily fired or detonated by a single person or a small number of people, the military can make good use of its available personnel: it can recruit selectively for a mission from the trained pool, lower personnel costs in the field, and reduce the chances that the mission will fail when essential personnel are lost.

The effect of the first two compensatory measures is to encumber communication between civilian and military leaders and/or leaders and individuals in physical control of the weapons.[6] The effect of all three measures is to increase the number of "weak links" in the system—that is, individuals who are "vulnerable to coercion, bribery, and blackmail" (Allison 2004: 91) and can seize control of the mission—and to reduce the number of people involved in a mission who can counter the unscrupulous, ingenuous, or incautious. The full control of a weapon could reside in a single individual or small groups of individuals, whose judgment, situational awareness, and motives will determine whether the weapon is used.

Of course, one way around these problems is to take subordinates "out of the decision." This is the intent of efforts to automate military responses by developing plans (and necessary technologies) to use weapons under preestablished sets of conditions. This would ensure that personnel know what to do when the threat is high and time is of the essence. Still, organizations have been known to shoehorn reality to fit the conditions that fall within the plans (on this, see Allison 1971; Lebow 1987). Militaries—like all organizations—take short cuts, develop rules of thumb, and devise procedures that "work" in some practical (military) sense but could have unwelcome consequences.[7] A classic example is the US Navy's "quarantine" of Cuba during the missile crisis. The Kennedy administration selected a blockade as the least provocative of military options, and then the administration sought to limit the provocativeness of the option by moving the quarantine line closer to Cuba (to give the Soviets additional time to reevaluate their strategy) and opting for shoot-to-disable over shoot-to-kill tactics against ships that violated the quarantine. At the same time, the Navy resisted pulling the line closer to Cuba (and thus Cuban air bases) and aggressively patrolled the seas to neutralize the Soviet submarine threat. These "sensible" military tactics could have undermined the "non-provocative" civilian strategy.

In sum, the intent of adversary offensive planning—motivated in part by the feared effectiveness of US defenses—is to shift the organizational default position from maintaining the status quo to ensuring that a weapon can or will be used under a wide variety of circumstances.[8] This is beneficial to the adversary if allowing it to accomplish its mission. It is disastrous if these measures also increase the likelihood that rogue-state weapons are fired due to miscalculation, misunderstanding, accident, or unauthorized seizures of command authority by low-level subordinates—as can occur given the likely deficiencies of a rogue-state arsenal. The nuclear forces of a rogue state—or any less developed country—might lack sophisticated and durable links between commanders and launch personnel for communicating when the country is under (nuclear) attack; sophisticated devices (so-called permissive action links, e.g. launch codes) that can prevent unauthorized weapons use; a satellite-based system to forewarn of an attack; an ability to protect intercontinental-range missiles in hardened, underground positions or through mobility; and advantages in size that permit the country to sustain a loss in missiles when attacked (see Feaver 1997; Seng 1997). The effects of each of these likely deficiencies will combine to increase the chances of an accidental launch and/or to create extraordinary pressure on a rogue state to fire its missiles or prepare them to be fired quickly and assuredly. Even factors that appear to militate against these undesirable consequences can increase their chances of occurring. For example, concealing weapons might reduce the urgency with which rogue states must respond to an (actual or impending) attack but might also limit a country's willingness to practice and openly plan for attack contingencies—preventing mishaps—out of fear of exposing plans and/or weapons locations to outsiders (Feaver 1997: 115–18). Thus, in an important sense, NMD advocates are betting that seemingly rational US policies will not be undermined by the compensatory actions of US nuclear adversaries, which they initiate merely to safeguard or extend their options (on this, see Glaser and Fetter 2005: 121).

That NMD advocates generally ignore these complications is somewhat ironic given the attention that policymakers have devoted to a related aspect of them—a "hand-off" of nuclear weapons by states to terrorist groups. The chances of a transfer seem greatest, too, when those leaders fear the impending collapse of their regimes and want to ensure that some entity is positioned to "act for" the regime with its demise. Yet autocratic leaders seem unlikely to engage in actions that amount to a hand-off of *political power* to potential competitors. Not only could the terror group decide when and how the weapon will be used—dragging the supplier (as the suspected source of the weapon) into a conflict—the group could blackmail the supplier to obtain concessions by threatening to use the weapon against some third party. Because these concessions could include the freedom to operate within and from the supplier country (maybe even the right to control government resources within that country), the threat to the sovereignty of the supplier is clear. Thus, the more likely possibility is that some loss of control will occur through internal failures of command within governments possessing these weapons.

The effects of pairing US offenses and defenses are not limited to rogue states and could include the destabilizing of the US deterrence relationship with nuclear powers, such as China. Because of US defenses, China has greater reason to take measures that can increase the chances of a nuclear exchange. Waiting out a US preemptive strike could leave China in the unenviable position of launching a "ragged" retaliatory attack against the US with what remains of Chinese nuclear forces, communication links, and command and control. From its impaired position, China might be unable to attack the defense in adequate numbers at its points of weakness, spread the defense (by hitting it with large numbers of missiles fired from different locations) to expose its vulnerabilities, nor guarantee a high probability of landing a sufficient number of warheads on target. As a precaution, then, China might plan to launch missiles rapidly with limited warning or evidence of an attack or China might decentralize its launch authority to guard against a knockout blow to the central command. These developments can make it more likely that China will respond precipitously to evidence of an attack. Although decentralizing authority could increase China's confidence that its forces will survive a US attack, decentralization could also increase the number of circumstances in which China will be too quick on the trigger if multiple persons acquire the capability to launch an attack. It is noteworthy, then, that China is moving to deploy a large number of mobile, solid-fuel missiles, which will be mated with their warheads (Chinese missiles currently remain unfueled and minus their warheads).[9] The relevant question is whether and when the units that physically control these mobile missiles can and would launch them.

The possibility of an unwelcome Chinese response is furthered by any number of factors that, from the Chinese perspective, magnify the offensive potential of US forces. These include the prominence in US nuclear planning of counter-force (e.g. military) targets and preemption, a grossly deficient Chinese missile warning system, a potential nuclear advantage that the US could obtain through a preventative or preemptive strike on China, attempts by defense proponents to vilify China (among other countries) to strengthen the case for missile defense, and US plans to construct a defense with capabilities that China views as unwarranted given the actual rogue-missile threat. The Chinese military already harbors deep concerns about US hegemony, the purposes of its alliances, and the ease with which the US employed offensive weapons and modern technology to defeat recent opponents (see, e.g. Shambaugh 1999/2000). Its offensive worries could increase as local rivals, Taiwan and Japan, are provided theater defenses (to protect against short-range missiles), if theater defenses are seen (correctly) to offer the US some future capability (when incorporated into a US NMD) to defend against Chinese long-range missiles (on this, see Glaser and Fetter 2001: 74; Wilkening 2000: 53–8), or if the US positions itself to destroy adversary missiles in their boost phase.[10]

This deterrence problem could also give Russia reason to resist important changes in its land-based missile force—specifically, de-alerting (e.g. separating warheads from missiles), downsizing, and de-MIRVing (i.e. replacing multiple independently targetable reentry vehicles—that is, multiple warheads on missiles—with single warheads). Such resistance would ensure that Russia retained a large, alert force of warheads. This is worrisome because Russia's high alert status and faltering attack warning system make it more likely that it will respond precipitously to erroneous warnings of attack. The chances of that occurring increase further because Russia relies on fixed-site, land-based missiles, and (at least for now) multiple warheads that reduce Russia's ability to withstand a US first strike. (Multiple warheads on land-based missiles are lucrative targets: one US warhead that destroys one Russian missile will destroy many Russian warheads.) These are significant problems for the US, if only because the possibility is great that *any* Russian attack on the US will involve a large number of missiles (Lewis *et al.* 1999: 38).[11]

The deterioration in Soviet nuclear attack capabilities and the vulnerability of the Chinese missile force, combined with improvements in the accuracy and responsiveness of US weapons systems, has an ominous consequence. It effectively transforms a US missile defense into a first-strike weapon—at least from the standpoint of the effected countries (see Lieber and Press 2006). From the Russian and Chinese perspective, a missile defense could give the US a capability to limit the damage from a counterstrike in a US preemptive or preventative attack. This reality is not lost on some proponents of missile defense. To their credit, cautious supporters of limited NMD acknowledge that the Russian or Chinese reaction to US defenses could leave the US in a net loss position—and, for that reason, they insist on *limited* US defenses and various reassurance measures to convince Russia and China that US goals remain modest (e.g. Daalder *et al.* 2000). But even these commendable measures downplay the rogue-state reaction to US defenses, the depth of Russian and Chinese skepticism toward US efforts, and the reality that US defenses can always improve as technology avails itself. Indeed, a US commitment not to upgrade defenses probably means little: the US formally committed under the ABM treaty not to develop the very systems that the US is now deploying.

The coercive use of nuclear weapons

NMD proponents argue that rogue-state threats to attack the US might leave it with no option but to capitulate to enemy demands. Thus, their ideas hark back to the Cold War era, when some strategists feared that the Soviets might gain a coercive edge with limited attacks that would leave the US without rational retaliatory options (Davis 1975; Schlesinger 1975). As before, the focus is on the risk-prone adversary that will resort to audacious—indeed, reckless—tactics to accomplish its political objectives. With closer scrutiny, it becomes apparent that the adversary's success at coercion might well hinge on: (a) retaliatory realities that give the US coercive advantages even without a defense and (b) the adversary's ability to capitalize on bargaining opportunities that a defense creates or allows.

US advantages without a defense

Any adversary that seeks coercive gains from attacking or threatening the US with one or more missiles invites a multitude of costs and confronts other constraints that strengthen the US bargaining position. These constraints reduce the effectiveness of the adversary's nuclear threats and the chances that nuclear weapons will be used. Importantly, these constraints work to the advantage of the US quite apart from whether or how well the US is protected by a missile defense system. The US advantages are as follows:

First, the benefits of attacking or threatening the US must be measured against costs that can be inflicted by an enormous US nuclear retaliatory force. Indeed, given the standard criticisms of US AD doctrine in the Cold War period, deterrence should be stronger now, in important respects, than it was in those years. In the 1980s, war-fighters pushed for missile-defense programs and offensive counter-force capabilities so as to "deny" the Soviet Union its gains from an attack when such deployments—under the most *optimistic* scenarios—would still have allowed the Soviet Union to deliver thousands of warheads against US targets. Now the US is positioned to inflict heretofore unprecedented levels of nuclear punishment upon considerably smaller rogue-state adversaries and to emerge from a nuclear conflict in a considerably stronger position than during the Cold War, in both absolute and relative terms. The US can render far greater destruction on these adversaries than they can inflict in return, and could use its superior might to eradicate adversary forces held in reserve. These twin realities place small nuclear adversaries in an extremely difficult military and political position. An adversary that uses its nuclear arsenal to maximum effect will have nothing in reserve for bargaining and will be exposed, potentially without reserves, to the full might of US retaliation;[12] conversely, an adversary that uses its missiles one (or a few) at a time will compromise their military effectiveness, risk the destruction of its reserve force, and undermine what is plausibly a key purpose of the "limited" attack, to avoid devastating US nuclear retaliation. It is odd, then, that the relative size of competing arsenals counts little in the thinking of some NMD proponents when it was a key concern of hawkish AD critics. Critics challenged the "absolutist" thinking of AD doctrine that rendered force "imbalances" inconsequential.[13]

Second, any nuclear attack on the US risks, at a minimum, an overwhelming conventional response and a concerted US effort to drive the rogue regime from power (not to mention, to hold its leaders accountable for "crimes against humanity"). That this threat may have been used effectively by the US to forestall an Iraqi chemical or biological weapons attack during Desert Storm (see CDI *et al.* 2001: 15; Freedman and Karsh 1993: 257; Posen 1997: 18–21) does not seem to impress NMD advocates. In congressional testimony, Deputy Secretary of Defense Wolfowitz (2001) asked rhetorically, "If Saddam Hussein had the ability to strike a Western capital with a nuclear weapon . . . would we really want our *only* option in such a crisis to be destroying Baghdad and its people [italics added]?" Ironically, despite decided US conventional military advantages in likely combat scenarios involving rogue states, and dedicated Cold War efforts to map the contingencies under which the US would respond to a Soviet *conventional* attack with *nuclear* weapons, NMD advocates so often appear to contemplate strategic nuclear options in a vacuum, devoid of conventional military responses.

Third, the US stake in local conflicts abroad is hardly inconsiderable. In the language of contemporary theory, the US has adopted a "costly signals" strategy that effectively bolsters US interests and makes it difficult for the US to back down in a conventional confrontation (on these signals, see Fearon 1994, 1997). For instance, through repeated public pronouncements that the US will use all means of force to defend South Korea, the US effectively "ties its hands"; the US risks damage to its international reputation and a public rebuke (audience costs), if it fails to stand by its commitment. Likewise, the sizeable conventional US presence in and around South Korea creates "sunk-cost" signals that convey to North Korea that conventional intervention on the Korean peninsula is, for the US, a cost-effective alternative. Admittedly, the relative stakes in bargaining might favor US adversaries. It seems reasonable to suppose, for example, that North Korea has a stronger desire for Korean unification than the US has to oppose it. But this need not mean that the US must blink first in a nuclear confrontation. The logic that has us discussing whether the "balance of interest" favors the US or North Korea assumes that some degree of rationality prevails in leadership decisions. If it does, North Korea's leaders must recognize that they have much to lose if a conflict "goes nuclear" or nuclear threats fail. The critical question, then, is not whether it is North Korea or the US that is most committed to its objectives but rather whether North Korea has anything to gain from striking the US and thereby placing at risk North Korea's cities, military capability, and leadership. In any case, if US adversaries still believe that they hold the higher ground in a confrontation, US defenses might be insufficient to correct the imbalance (as discussed in the section "Extended deterrence and defense" on p. 94).

Fourth, the structure of conflict works against US adversaries. To say that rogue adversaries can coerce the US with a threat of nuclear attack is to *stand on its head* the logic with which Cold War critics assailed AD doctrine. Remember, critics maintained that deterrence was threatened because it was irrational for the US to respond to a Soviet attack: in essence, the Soviet Union could use "limited" attacks to constrain US options, shifting the burden of triggering a suicidal exchange onto the US. But it is now the rogue-state challenger that finds its options constrained: if the US were to stand behind its conventional commitments in Korea and elsewhere, the challenger must bear the burden of taking the conflict to the next level by triggering a fatal exchange. It was just this strategy of shifting the burden to the attacker that served US purposes in Cold War crises. In launching an around-the-clock airlift to Berlin in 1948 in response to the Soviet blockade on land traffic into the city, the US forced the Soviets to take the next move, if it dared: the Soviets would have to shoot down US aircraft to enforce the blockade of the city. In opting for the quarantine of Cuba in 1962, the US forced the Soviet Union again to initiate the provocative countermove: Soviet vessels would have to shoot their way through the US naval blockade to resupply Soviet forces in Cuba. In both instances, US leaders relied cautiously upon the risk-averse impulses of Soviet leaders. History judges these decisions kindly.

US disadvantages with a defense

This does not mean that missile defenses would provide the US with no bargaining advantages. NMD advocates have a point when they argue that defenses can bolster the US bargaining position in a conflict in which the strength of US forces, US stakes in a conflict, and the logic of the situation little affect rogue-state calculations. If these adversaries know they can do little harm by attacking the US, given its missile defense, the effect is to weaken the credibility of those parties' threats to use force. But US defenses could also give these adversaries a bargaining edge, if the speculations of Cold War strategists hold true. Following Schelling (1960,

1966), strategists recognized that the disadvantaged party (then, the US) could obtain coercive leverage through a "competition in risk taking" or a "threat that leaves something to chance."[14] These strategists did not necessarily believe that the US could fight and win a nuclear war,[15] but they did believe that the Soviet Union might back down if the US sent the right mix of messages, with nuclear strikes tailored to signal both resolve and restraint. Their thoughts fed the emphasis on "limited nuclear options" in US nuclear doctrine of the early 1970s; in the present context, these same ideas suggest, for a number of reasons, that the adversary is favorably positioned.

First, US defenses can become coercive targets for the reckless and risk-prone adversary (envisioned by US policymakers) that possesses a survivable retaliatory force. If Cold War strategists were correct that the US could obtain a coercive edge by staging (symbolic) nuclear attacks that were limited by means, effect, and target, an adversary might achieve similar advantages by attacking defended targets. These attacks would allow the adversary, in theory, to show that it means business, without generating (counterproductive) collateral damage. Even if the coercive effort failed— and the US were to respond with overwhelming nuclear force—there is also a possibility that US defenses will not deflect the adversary's symbolic attack, which could result in large numbers of US fatalities. To thwart these attacks, the US has an impossible decision: it cannot advertise the effectiveness of its defense since this risks attacks on US targets; it cannot confess the weakness of its defense since this hurts US bargaining leverage.

Second, a defense could place the US at a coercive disadvantage against the same reckless adversary, before any weapons are fired, by creating the perception that the US lacks resolve. Importing the logic of the Cold War deterrence literature, a missile defense could weaken the US bargaining position if it suggests to opponents that the US values the lives of its citizens more than it values the stakes at issue. By seeking to protect the lives of Americans, the US is unintentionally surrendering leverage by conveying to an opponent that it might not want to stand by a commitment if Americans would be hurt in consequence. Indeed, it could suggest that the US is unwilling to retaliate if attacked because it cares about sacrificing lives, at home or abroad, and will not retaliate lest it take innocent lives or invite retribution. At the very least, protecting US citizens indicates to an opponent that the US expects an attack, which means then that the US cannot call the opponent's bluff. The relevant metaphor is an unarmed, Western sheriff who walks confidently and deliberately toward a pistol pointed in his direction. By revealing no apparent concern for his personal safety, the sheriff seeks to convey the message, "you won't really shoot." The message is that the circumstances speak for themselves; the logic of the situation promises that the gun-wielder will not act on their threat. On this point, Cold War-era strategists acknowledged that efforts by the US to reduce its exposure to the devastating consequences of a nuclear war could negatively affect the US bargaining position.

Third, against an adversary with strong security concerns, a defense could place the US at a coercive disadvantage before any weapons are fired by fostering the impression that the US has aggressive intent. Efforts by the US to defend itself—when the adversary is unthreatening from its perspective—could be interpreted by the adversary as an offensive gesture. This is the power of "reverse imaging." National leaders no less than other people reserve the most favorable judgments for their own behavior and less favorable judgments for adversaries. Given a preexisting negative view, it is easy for a weaker party, then, to read offensive motives into an opponent's defensive acquisitions. If a leader *knows* that its intentions are "pure," and has reasons to suspect that US intentions are not (e.g. the size and capabilities of the US offensive arsenal and the US rhetorical pursuit of a preemptive strategy), that leader can only conclude that the US seeks to protect itself because the US plans to attack. Convinced that it is about to be attacked, the weaker party has little to lose from an aggressive, coercive posture. Indeed, it might be able to use its strategic disadvantage to obtain a coercive edge. A cornered bank robber with no option but to hold a gun to a hostage's head can still force the police to retreat. The threat to the hostage is given credibility by the very precariousness of the robber's situation.

The obvious response to this logic is that a preemptive strike against a vastly superior foe would be suicide, yet this response is proscribed by the logic that motivates this discussion. If the adversary is unwilling to strike, then concerns about coercive advantages are misplaced; if it is willing to strike, then preemption could make sense to the adversary. More consideration is due a second counter-argument: that adversary coercive advantages can be countered with a strong defense, indeed, that there is no coercive counter to a perfect defense. From the standpoint of bargaining and deterrence, though, the outcome is not that clear cut. First, the adversary can redirect its fire to third countries (as discussed, as well, in the section that starts on p. 94). In the blackmail game, if you cannot coerce someone directly, you can do so indirectly by threatening their family. Second, there is no such thing as a "perfect defense": any residual threat (or US doubt about the effectiveness of the defense) is a basis for coercion. Would an adversary that believes that it has a coercive advantage with four warheads (pitted against thousands in the US arsenal) believe that it has a fundamentally smaller advantage with, for instance, a one-in-five chance of landing one warhead on US soil? After all, no larger purpose appears to be sacrificed when one city is destroyed instead of four. For its part, would the US play Russian roulette by accepting a one-in-five chance of a US city being destroyed? Is this a reasonable risk against an adversary that NMD advocates believe is prone to risk taking? To the contrary, even the architect of the demanding AD standard professed to be squeamish about the prospect of a single Soviet warhead landing on US soil: in Robert McNamara's view, no president would "consciously sacrifice an important part of our population or our land and place it in great jeopardy to a strike by Soviet strategic forces, whether it be one city, or two cities, or three cities" (quoted in Lebow and Stein 1995: 170).

Again, speculation that adversaries might use their weapons audaciously is not meant here to suggest that adversaries are reckless nor coldly conniving, that government leaders have the presence and cognitive skills to read messages as they were intended in or outside the fog of war (see, e.g. Jervis 1982/3, 1989), that threats deter escalation rather than provoke it (on this, see Jervis 1976; Lebow and Stein 1989), nor that bargaining advantages can be obtained from the threatened or actual use of nuclear weapons. It is offered, instead, to show that arguments made by NMD advocates support a defense policy that, *at best*, complements retaliatory deterrence and might weaken it. In a world in which the US must rely upon missile defense to compensate for the coercive limits of overwhelming US power, the US might remain disadvantaged by any defense that is imperfect or that is seen as such, by US policymakers.[16]

The offense-defense balance

Many NMD advocates hope to avoid nuclear relationships, with rogue states, that are based upon principles of AD that keep US cities at risk. For these hopes to be realized, program advocates must set realistic goals for system performance that, if met, will free the US from deterrence relationships. Should these goals prove unobtainable, the US must be better off with available defenses than without them. The question, then, is whether the US is fundamentally advantaged or disadvantaged in an arms competition that pits US defenses against the plausible offensive capabilities of potential nuclear adversaries. The answer is not as favorable to the US as NMD advocates assume. Despite the performance claims and aspirations of program supporters, a simple statistical analysis reveals that, by deploying a missile defense, the US could spark a competition that plays to the advantages of adversary offenses.

Program performance

Program defenders argue that a complex of satellite and ground-based sensors, warning, tracking, and targeting systems and the modesty of US program objectives will ensure program success. They insist that, unlike the schemes of the Reagan administration, NMD will employ proven technology (e.g. hit-to-kill interceptors instead of satellite-based, X-ray lasers) to defeat a small threat (i.e. a handful of single-warhead missiles rather than thousands of Soviet warheads). The optimism of NMD advocates reflects their faith that defensive technology can be made to perform and that offensive improvements and advantages will not erase defensive gains.

Critics respond by noting that defensive systems are vulnerable to a variety of offensive strategies and countermeasures. First, the offense can design its attack to capitalize on defensive vulnerabilities while the defense must prepare for attack across the full range of offensive options. In addition to a coordinated or surprise attack, these options include low-trajectory missiles fired from ships, missiles bearing biological weapons "bomblets" (or submunitions) in the hundreds that are not easily destroyed, or non-missile nuclear threats, e.g. the suitcase bomb. Second, the offense can rely on weapon quantities to compensate for qualitative disadvantages. An attacker can capitalize on economies of scale to construct larger numbers of warheads, while the defense may be limited in the number of interceptors that can be directed to any warhead. Third, the offense has available to it countless technological options against which the defense must prepare. Rogue states could develop a variety of countermeasures (Lewis *et al.* 1999: 39), e.g. warhead decoys, or acquire them from more-advanced nuclear countries, e.g. China.[17] Program defenders feel confident, however, that the threat from rogue states, in its sophistication and quantities, will not severely challenge US defenses. They acknowledge that adversaries can theoretically counter a missile defense but they doubt that the required technologies will be available to cash-strapped, politically isolated, and technologically challenged countries such as North Korea and Iran. For that reason, designers hope for a defense with a ninety-five percent effectiveness against an attack by warheads apparently numbering in the tens.

Whether a ninety-five percent "kill ratio" is a realistic objective is hard to know for certain but it is, without question, an ambitious objective because of the difficulty, if not impossibility, of knowing whether the standard has been met. To establish ninety-five percent effectiveness requires an extremely large number of consecutively successful tests on an operational system that must perform a complex and coordinated set of tasks. Thus, the ineffectiveness of a system is a *multiplicative* function of the failure of the system to perform each of its specific tasks. Such interdependence in system performance means that as the number of components increase, the chances that the system will successfully accomplish its mission drop rapidly, and then precipitously. A system with three, seven, and ten components, each one of which has a ninety percent chance of working as planned, stands around a three-in-four, one-in-two, and one-in-three chance, respectively, of successfully accomplishing its mission. (If the components are interdependent, so that reliability problems are contagious, the chances of mission success are still lower.) Thus, when program advocates discount the importance of "simple glitches" that do not represent "major deficiencies in the system," they miss the point that complex technologies are no more reliable than the probability that a simple glitch will prevent the system from performing as required.

Beyond this, high levels of effectiveness have not been demonstrated in developmental tests of system components, let alone tests of an operational system under wartime conditions. The missile development program has been plagued by test failures that have involved even "non-problematic" technologies and less-challenging aspects of the intercept mission such as the missile launch. Even then, it is appropriate to ask what it means for a system to pass a developmental test. These tests have not served traditionally as a "gatekeeper" to prevent problematic weapons from entering production and eventual deployment. Tests have typically not been designed to expose the limits of a technology but to play to its strengths. For instance, the interception task in tests of missile defenses was eased when devices were placed on mock warheads to signal defense interceptors and when decoy warheads were made to *appear* distinct from "actual" warheads to help target discrimination. Because tests are not intended to make the job harder for the defense but are intended, as a learning exercise (to "demonstrate" capabilities), the meaning of the results is in the eye of the beholder—here, program officials, who are inclined to view test data as revelatory, and setbacks as challenges (Lebovic 1996: 110–15). Even with demanding tests, doubts about system effectiveness must remain. Realistic tests, in which missiles equipped with actual adversary countermeasures are fired from hostile territory, are impossible. Absent realistic tests, the US must deploy a system that will receive its first true test when it is needed: "there will be no opportunity to learn on the job."[18] Without these obstacles to testing, the challenge would still be great: ninety-five percent effectiveness is "rarely—if ever—achieved by a military weapons system, even after years of use" (Lewis *et al.* 1999: 126).

A simulation

As the following analysis shows, even a high rate of effectiveness might not free the US from deterrence relationships; and lesser effectiveness quickly ensures a defense that is undoubtedly tied to the principles of retaliatory deterrence and can leave the US at a net disadvantage. These conclusions draw from statistical calculations, based on three assumptions. These are: (a) that the effectiveness of the defense can be expressed by the probability ("kill ratio") that it will destroy any attacking warhead; (b) that the number of attacking warheads ("attack force size") can change as an attacker enlarges the size of its attack force or devotes more or fewer weapons to an attack; and (c) that the attacker (defender) seeks a capability to hit (protect), with some probability, some number of targets that is less than or equal to the number of warheads.[19]

Figure 4.1 reveals the effectiveness of different size arsenals against defenses of varying strength. It expresses this effectiveness as the probability that at least two, four, six, and eight warheads (in 1a–1d, respectively) survive the US defense. In all four graphs, the x-axis marks the size of the attack force. The lines in each graph portray the defensive kill ratio, the percentage of incoming warheads that the defense can disable, deflect, or destroy: a kill ratio of 0.10 indicates that only one in ten incoming warheads can be destroyed, kill ratios above 0.90 express the high effectiveness levels to which defense proponents aspire. The y-axis represents warhead survival probabilities (by definition, these vary between 0 and 1.00). In Figure 4.1a, for instance, a point on a line represents the one-warhead survival probability at a given attack force size with a given kill ratio. The graphs can thus be understood to represent the competition between defensive quality (the kill ratio) and offensive quantity (the attack force size).

Figure 4.1a shows, then, that an attacker desiring modest success (that is, that at least one warhead will penetrate the defense) can achieve that goal with high probabilities, at relatively small force levels, when the defensive kill ratio is low. With kill ratios around 0.50 (i.e. the defense has a 50:50 chance of defeating an incoming warhead), the probability that at least one of three warheads will survive the defense is roughly ninety percent. Against more effective defenses, survival probabilities still rise with increases in the number of warheads launched. Note that the one-warhead survival probability is near or above ninety percent for a six-warhead launch when the kill ratio is 0.60, a seven-warhead launch when the kill ratio is 0.70, and a ten-warhead launch when the kill ratio is 0.80. When the kill ratio rises to ninety-five percent, warhead survival probabilities drop substantially and the one-warhead survival probability reaches ninety percent only when the attack force is so large that it is not represented on the graph.

These four graphs have many important implications for deterrence and defense. First, deterrence remains at issue should system performance goals be met. Even pitting a small threat (four warheads) against the defensive capability to which NMD advocates aspire (kill ratios = 0.95) produces a sobering result, a nearly one in five chance that a warhead will land on US soil.[20] With a dozen warheads, the attacker's chances of landing a warhead approach 50:50. Although this destructiveness falls well below even those low levels of retaliatory capability that "minimum deterrent" advocates argued would deter the Soviet Union (e.g. one Trident submarine), a one in five chance of destroying a US city cannot easily be dismissed. Outside the world of nuclear strategy, that is an unacceptable risk.

Second, defense demands excellence. Comparing the performance of the strongest defenses (kill ratio ≥ 0.95) with weaker ones exposes a startling reality of offensive–defensive competition: it is a sharpshooter's game; the 0.95 effectiveness criteria used by defense proponents is a statistical

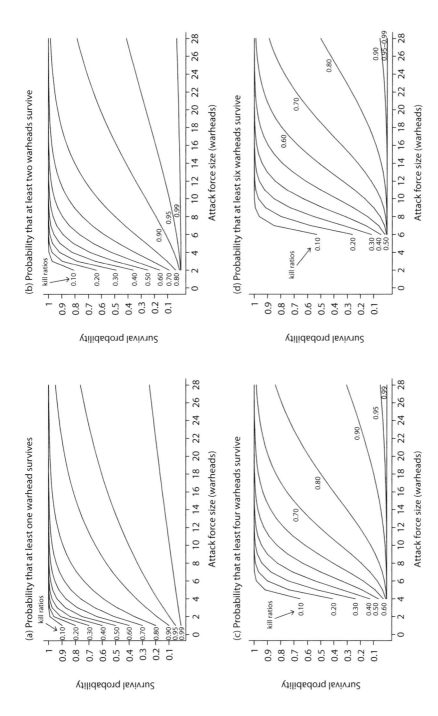

Figure 4.1 Warhead survival probabilities given different force sizes and kill ratios

necessity. Below such high levels of effectiveness, warheads too easily penetrate the defense and the defense suffers enormously if the attacker augments its attack force. With ninety-five percent effectiveness, one-fourth of the time, one in six warheads will penetrate the defense; one in twenty times, two of these six warheads will penetrate the defense. With ninety percent effectiveness, a six-warhead force will deliver a warhead almost fifty percent of the time and, two warheads, over ten percent of the time. If defensive capability falls below 0.90 effectiveness, the marginal impact is large. A kill ratio that drops to 0.80 will allow one of six warheads to penetrate nearly three-quarters of the time and two of six warheads to penetrate over a third of the time. The strongest defenses also yield disproportionate protection against more severe contingencies. Observe, for instance, that there is over a one in three chance that six of twelve warheads will penetrate a defense with a kill ratio of 0.60 (Figure 4.1d), whereas the probability that even four warheads will penetrate a defense with 0.95 effectiveness is close to zero (Figure 4.1c).

Third, under some conditions, incremental improvements in weak or moderately effective defenses produce small payoffs compared to the marginal returns that are realized by building on an already strong defense. When the number of surviving warheads that the attacker is seeking is small relative to the size of its attack force, defensive performance improves rapidly only for the most capable of defenses. With an attack force size of six warheads, a 0.50 drop in the one-warhead survival probability is realized by increasing the defensive kill ratio from 0 to 0.90; an improvement almost equal to that is realized by increasing the kill ratio from 0.90 to 0.99.

Fourth, any battle between offense and defense in which the defense manages to "keep up" with offensive improvements can be a battle "lost" by the defense: the rules of mathematics promise continuing *relative* success, against weak to moderate defenses, to an attacker that can enlarge its forces. This can be seen more clearly with the aid of Figure 4.2. Figure 4.2a reveals that when defense effectiveness falls below 0.60 (or 0.70, with an attack force greater than six warheads), the survival of one-quarter of the attacking force is virtually assured: against these weaker defenses, the probability that a quarter of the attack force will survive rapidly approaches and then surpasses ninety percent, as the size of the attack force increases. A more ambitious attacker must count on a weaker defense. As Figure 4.2b shows, an attacker that wants half of its force to survive must hope for kill ratios around 0.40. Of course, when an attacker maintains the same *relative* success with a larger attack force, a larger *absolute* number of warheads penetrate the defense.

Fifth, a defense that "looks stronger than it is" invites calamity, despite its apparent deterrence benefits. Weak to moderately defended targets could

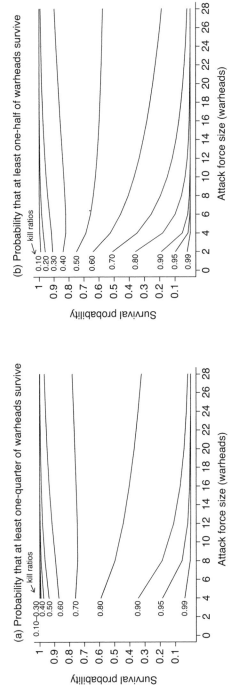

Figure 4.2 Force fraction survival probabilities with different kill ratios

suffer severely if the defender increases the size of its attack force or seeks to degrade the defense's performance to offset imagined defensive advantages. In fact, small growth in an attack force combined with modest reductions in defensive performance could have large effects on warhead survival probabilities. For instance, if the attack force grows from four to eight warheads, while defensive effectiveness drops from ninety to eighty percent, the two-warhead survival probability rises from five percent to fifty percent. If effectiveness drops to fifty percent, the effects include a ninety-six percent chance that two, and a sixty-four percent chance that four, of the eight warheads will penetrate the defense.

The worst-case assumptions of larger nuclear adversaries are worth considering, too. That those defenses might only be effective against a portion of the Russian or Chinese force misses the point. For instance, with worst-case Chinese assumptions about the capabilities of a "limited" US NMD system (of say, 100 interceptors), it can be supposed from these graphs that China will need a force many times larger than its current force to maintain even its existing penetration capability, and a still larger force to counter the vulnerability of its existing force to US preemption, to guard against rapid expansions in US defenses, and to ensure conclusively and communicate unambiguously that Chinese missiles can penetrate those defenses.[21] Sizable increases in the Chinese arsenal could spur an arms race with still other countries (e.g. India),[22] if not vitiate efforts at US and Russian arms control. From the Russian perspective, every missile against which the US can defend is effectively a missile that cannot be removed from those arsenals by negotiation or that can be removed only with asymmetrical US concessions.[23]

Thus, limited US defenses could trigger undesirable responses from US nuclear adversaries, both big and small. Indeed, an arms race of quite modest proportions could hurt the US if its defenses fail to meet the most demanding effectiveness standards. Moreover, even effective defenses are unlikely to supplant US reliance on retaliatory deterrence, and can render the US worse off, in the long term, than it would have been without a defense. This could occur if the adversary adds but a few warheads to its arsenal and degrades the defense to levels at which it can still destroy a substantial portion, e.g. seventy percent, of attacking warheads. The evidence suggests that, contrary to the claims of program proponents, a little defense is *not* better than nothing. Even defenses that will stop most incoming warheads can put the US at a disadvantage.

Extended deterrence and defense

The quantitative analysis is somewhat static in that it assumes a uniform defense. Missile defenses will be uneven for the foreseeable future, giving would-be attackers an incentive to shift targets to avoid effective defenses.

Although US targets could have high value to a hypothetical attacker, the low likelihood of destroying a US target could dissuade an attacker from launching its missiles against targets within the US. Even the attacker that believes the US is the root of all evil might not attack the US directly if it could thwart a planned offensive. Unfortunately, the attacker that NMD proponents envision—one bent on producing destruction at any cost—is presumably open to a range of possible targets and could strike out against perceived US surrogates and interests in other parts of the world. At a minimum, the adversary could threaten US allies to reap concessions from the US. The upshot is that a paradox of missile defense is that the defense of some targets (the US) makes other targets less secure.

This side of the global security dilemma is not given much attention by NMD advocates or, for that matter, by international politics scholars. As generally conceived, the security dilemma has a powerfully simple logic: when a state arms in defense, other states become less secure (Jervis 1978). That is, defensive actions create offensive threats. The dilemma is rooted in the relative nature of military capability: increasing the capability of state A effectively decreases the capability of state B. But unrecognized (or, at least, under-recognized) in this dilemma is that the defense can feed offensive fears by redirecting the offensive threat. Just as squeezing any part of a balloon increases the pressure on its other surfaces, defending a target transfers threat to alternative targets (increasing the probability that they will be hit). This problem has long bedeviled counter-terror strategies (see Chapter 6), as the bombing in Atlanta during the 1996 Olympics demonstrates. Enhanced security precautions taken at Olympic events and in the Olympic village only worsened the danger to fans and athletes who gathered elsewhere. The insecurity of less-protected targets grows further if the attacker initially upgrades its capability to meet the challenge posed by attacking better defended targets. The problem can be understood more precisely by returning to the statistical analysis and exploring its implications.

Extending the simulation

Just how insecure less-defended targets become depends again on the size of the attack force and the capability of the defense. Figure 4.3 illustrates. It presents the attacker's incentive to shift targets under two of the conditions found in Figure 4.1: specifically, when the attacker possesses four and eight warheads (Figure 4.3a and 4.3b, respectively). In this figure, however, the defense kill ratio appears on the x-axis (it is now a continuous variable) and the lines in the graph indicate the survival probabilities for some (non-zero) number of surviving warheads.

Figure 4.3a shows an attacker, with four warheads, that has three simple choices. First, it can hold to its plan to attack a fairly well-defended target

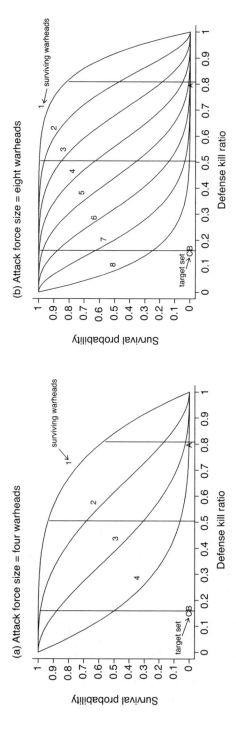

Figure 4.3 Incentive to shift to less-defended targets given different force sizes and kill ratios

set, A (kill ratio = 0.80). If it does, it can count on a fifty-nine percent chance that one warhead will survive the attack and an eighteen percent chance that two warheads will penetrate the defense. Second, it can redirect its attack to a moderately defended target set, B (kill ratio = 0.50). If it does, it can plan on a ninety-four percent chance that one warhead, a sixty-nine percent chance that two warheads, and a thirty-one percent chance that three warheads will penetrate the defense. Third, it can attack a set of lightly defended targets, C (kill ratio = 0.15). Against these targets, it can count on the near certainty that one warhead, the strong likelihood (a 0.89 probability) that three warheads, and a slightly better-than-even chance that all four warheads will survive the attack. All other things being equal, then, the attacker has a strong incentive to shift its attack to the less-defended target set. In fact, large benefits can be realized by an attacker willing to shift its attack to slightly less-defended targets (a kill ratio of 0.50 rather than 0.80): to an attacker, hitting B is highly preferable to hitting A, and hitting C offers a much larger payoff than does hitting either A or B. Among other advantages, striking B instead of A offers twice the amount of warhead penetration (two warheads versus one), with a higher survival probability; it also improves greatly the chance that a single warhead will survive the defense. Therefore, the defense of A has strongly increased the vulnerability of alternative targets, B and C. If the attacker prefers to destroy A over B and C, the latter are rendered vulnerable, nonetheless, if the damage inflicted upon them compensates for their relatively low value.

The vulnerability of these targets only increases if the adversary increases the size of its attack force. Comparing Figures 4.3a and 4.3b establishes that doubling the size of the attack force, adding but four warheads to a four-warhead attack force, greatly increases the attacker's options: a large number of better defended targets can be hit now with greater certainty. Against A, the attacker can now plan for an eighty-three percent chance that one warhead, and a fifty percent chance that two warheads, will survive the defense. But, against B, the attacker can plan for a near 100 percent chance that one warhead, a ninety-six percent chance that two warheads, and a sixty-four percent chance that four warheads will overcome the defense. Against C, the attacker can anticipate a near 100 percent chance that *half* of its eight warheads will survive the defense—in fact, a sixty-six percent chance that seven of eight warheads will penetrate the defense.

For that matter, the attacker's incentive to shift to less-defended targets can increase enormously when the attacker adds but one warhead to its strike force. Figure 4.4 displays the marginal effect on the one, two, four, and six-warhead survival probability (Figure 4.4a, b, c, and d, respectively) when a single warhead is added to the attack force. It shows that adding one warhead can greatly increase the chances that the defender's "worst case" will be realized, that is, that most or all warheads penetrate the defense. (Because the possibility that all warheads survive is not considered in the figure, the worst case, here, is that all but one warhead survives.) Note that in Figure 4.4a–d, the highest marginal improvements in survival for one, two, four, and six warheads occur when force sizes increase from one to two, two to three, four to five, and six to seven warheads, respectively, and that the marginal improvement declines as the number of surviving warheads constitutes a smaller percentage of the attack force. Note, too, that this increasing capability to inflict the worst case on the defender is realized against *relatively weak* defenses (and, to a lesser extent, against moderate defenses).[24] These marginal effects are not to be taken lightly: Figure 4.1 shows that, with low kill ratios, these effects are sufficient to push the warhead survival probabilities into the upper range. In contrast, note that it is hard to boost survivability against strong defenses.

In sum, small force size improvements make life significantly more difficult for weakly or moderately defended targets, not targets that are defended at the high levels to which US policymakers aspire in homeland defense. It is apparent that *only* one warhead added to an attacker's arsenal can significantly increase the probability of a catastrophic result for less-defended targets. These effects are worsened when cumulative. Adding one warhead to a two-warhead force increases the two-warhead survival probability against a moderate defense (kill ratio = 0.50) by twenty-five percent, adding another warhead increases two-warhead survivability by nineteen percent, adding still another warhead increases two-warhead survivability by thirteen percent. Under these same conditions, the vulnerability of heavily defended targets increases only slightly.[25]

Implications

When all is said and done, the act of defending a target can be viewed as an act of "defection" (in game theory) or "buck passing" (in balancing theory) because defending a target essentially passes the burden of protection on to less-protected targets. But it does more than that; it actually renders the latter more insecure. As the scope of available targets contracts, less-defended targets carry a higher security burden, evermore the victims of the world's redirected hostility. The question, then, is "will deterrence be sufficient to compensate?"

The difficulty of extending deterrence to non-US targets was well appreciated by Cold War-era deterrence theorists. Because an all-out nuclear war between the superpowers would be suicidal, it was feared that the Soviets might be able to attack US allies with impunity. Consequently, strategic theorists sought various means to extend deterrence to third countries to prevent them from becoming "decoupled" from the US retaliatory commitment, should the Soviets doubt that the US would aid these countries if they were attacked. In doing so, however, these theorists (or, at least, their

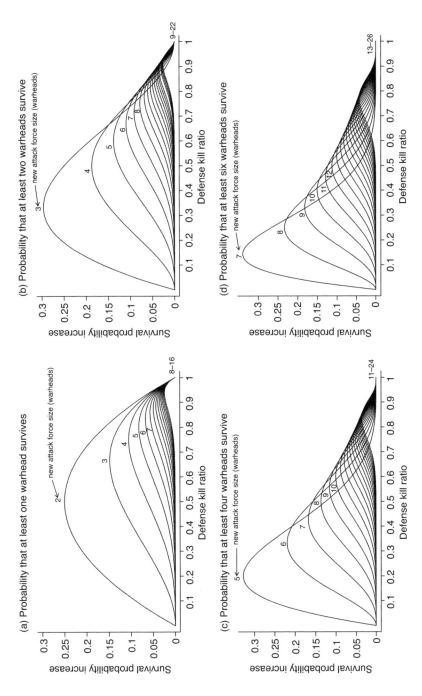

(a) Probability that at least one warhead survives

(b) Probability that at least two warheads survive

(c) Probability that at least four warheads survive

(d) Probability that at least six warheads survive

Figure 4.4 Effects on survival probabilities of adding one warhead to the attack force

critics) appreciated that actions taken to bolster US commitments abroad might weaken them: for instance, the deploying of US conventional or theater nuclear forces in Europe might convey a US reluctance to retaliate fully against Soviet transgressions there, making them more likely to occur.

Likewise, providing missile defenses to third parties could undermine deterrence by sending the wrong signal to adversaries. By strengthening regional defenses, the US might appear unwilling to extend deterrence to effected countries. Or, as some Europeans fear, the US might create a "self-fulfilling prophesy: by admitting concerns that traditional deterrence might fail, the US pursuit of NMD will help to ensure that it will fail" (Gordon 2001: 25). NMD advocates see it differently though. For them, the danger of decoupling is a major reason for NMD (see, e.g. Hadley 2000; Slocombe 2000). As one Clinton administration official asked, "Why would the United States be a better ally if it were vulnerable to North Korean missiles?" (quoted in Daalder *et al.* 2000: 16). NMD advocates argue, then, that a well-defended US can use its nuclear (and conventional) might to offset the vulnerability of countries abroad so that the expected costs of upholding commitments abroad are brought into balance with the benefits. In fact, these advocates argue that strengthening local defenses—providing defensive technology to third countries—is a means by which the US can signal its intent to extend deterrence. But, if missile defenses are to permit the US to extend deterrence, a number of conditions must hold. These follow directly from the paradox that makes US allies more desirable targets as US targets become less vulnerable.

First, US retaliatory threats must be effective against an adversary with a now stronger incentive to attack US allies. Whereas NMD advocates argue elsewhere that rogue states might attack US territory despite possible US retaliation, they argue here that these states can be deterred by what can only be a lower probability that the US will respond to an attack on a non-US target. Indeed, the US must appear willing to use its nuclear weapons against a rogue state when *American* lives are not directly at stake, when the US faces the undesirable prospect of punishing a dastardly deed by committing another one (incinerating innocent civilians), and when the US might have to retaliate after the shock from the initial attack has subsided and retaliation cannot be justified militarily nor politically, perhaps because all adversary missiles have been fired or the rogue leadership is no longer in power.

Second, the credibility of US retaliatory threats must offset the increased incentive that the adversary now has to attack US allies, despite the probability that one or more adversary warheads will land on US soil. With or without a US NMD, then, the US proclivity to intervene abroad will most certainly reflect the US commitment to the goals at hand, US beliefs about whether and when a foreign adversary will strike the US with nuclear weapons, and the US willingness to persevere in the face of—perhaps, because of—that missile threat. Hence, when the soon-to-be Deputy National Security Advisor (and eventual National Security Advisor; Hadley 2000: 102–3) lamented that the threat of attack by ballistic missiles armed with *conventional* warheads might be enough to deter US intervention on behalf of its allies, he described a problem that US NMD cannot cure.

Third, the US must accept the political and economic burdens that go with defending US allies that are rendered more vulnerable to attack. In supplying missile defenses, the US must consider the financial cost and political value of defending countries when vital US interests are not at stake, when defending some countries may increase the pressure on the US to defend still others, when these defenses might be taken for an alliance by friends and adversaries (leading either or both to act provocatively), and when the US might face strong pressure to extend *deterrence* when local defenses are inadequate or prohibitively costly.

Finally, the deterrent value added to a regional target by a US commitment to defend it must be larger than the added incentive that an attacker has to strike that target by merit of that commitment. Put simply, if the adversary is drawn to "surrogate" targets, any US commitment to defend a country increases the threat to it. It is possible, for instance, that Patriot missiles made Israel and Riyadh more attractive targets to Iraq in the first Gulf War than they would otherwise have been.

In all, then, missile defenses invite, and may exacerbate, traditional deterrence dilemmas. They provoke the attacker to look for openings that could leave the US without desirable responses. In the final analysis, a US NMD might not increase the credibility of US retaliatory threats sufficiently to offset the increased incentive that the attacker has to strike US allies by virtue of those defenses.

Conclusions

While old deterrence arguments lie dormant, strategic arsenals and planning do not. A Cold War legacy remains in thousands of strategic missiles that still point at their targets and an even larger number of warheads that are stored for use. Some problems have gotten worse: Russia's warning systems are deteriorating, its strategic arsenal is in decay, its missiles remain on alert for a quick response to warnings of attack, and its instability and economic distress have fed grave concern over the custody of Russian nuclear weapons and materials. Because such problems lie in the complicated, turbulent, and disappointing world of economics, politics, and diplomacy, they are not addressed with the enthusiasm that greets problems that can be solved, so it seems, with technology. In pinning their hopes to missile defense, however, advocates pair a hypothetical problem with an expensive solution that might fall short and could aggravate underlying security problems. They fail to appreciate the full extent to which the "cure," here, is worse than the disease, if opponents counter a defense by relaxing weapons controls or launch criteria, by capitalizing on bargaining leverage that US defenses afford, by channeling the arms competition into weapons against which defenses cannot defend, by seeking quantitative solutions to offset defensive qualitative advantages, or by increasing the threat to US allies. Missile defenses will not supplant deterrence, then, they will weaken it.

To be fair, even NMD enthusiasts might acknowledge, if pressed, that deterrence has a place in combating new threats; their faith that these threats could be countered by a defense would probably diminish significantly if not for the presence of the US retaliatory arsenal. In that sense, enthusiasts validate the beliefs of more cautious NMD advocates who try to balance deterrence and defense. But striking that balance is no small achievement. Indeed, cautious supporters should reconsider their qualified support for missile defenses if they are assuming that the liabilities of NMD can be mitigated simply through a reassurance policy meant to assuage those who question US motives. The principles of Cold War deterrence relied, not on the intentions of protagonists, but rather on their capabilities. It was for that reason that the US and Russia moved away from MIRVs and land-based missiles in the Strategic Arms Reductions Treaty (START) I and II in the 1990s. These systems were regarded as weapons that could *create intent* in the form of an incentive to preempt in times of crisis; that US or Soviet (Russian) intentions were *fundamentally* benign was beside the point.[26] These advocates should also reconsider their support if they are betting that prudence will prevail in NMD policy—that technological success will not feed demands for better and more complete defenses and that defenses will not rise, as a consequence, to their level of incompetence and produce uncontrollable problems. The reality is that the Bush administration paired its enthusiasm for missile defense with an expressed distaste for arms control agreements, conflated Russian and Chinese resignation toward NMD with acceptance of it, implied that the current immaturity of technology poses the greatest impediment to an upgraded defense, refused to set upward limits on the numbers of interceptors that a defense system will employ, and considered incorporating a variety of systems, developed with different purposes (e.g. theater defense), into a national defense system (O' Hanlon 2001).

A missile defense could plausibly be constructed with proper sensitivity shown to the limits of defensive technology and its confounding political and strategic effects. But the challenges are great; and any defense, including reputedly more benign systems, e.g. (earth-based) boost-phase defenses, might always be a blunt instrument in the hands of government. Whatever the actual effectiveness of these systems,[27] troublesome questions remain about how they will be linked to other forms of missile defense, how US adversaries will view those systems, and when and where they will be deployed. If deployed in a crisis, for instance, boost-phase defenses could precipitate the attacks that those defenses were meant to prevent and deflect.

In the end, a missile attack on the US by a rogue nuclear state remains a frightening possibility. The rationality that prevails in time of peace can degenerate in time of war; and history offers proof of megalomaniacal leaders who willingly accept huge costs to serve unworthy, and ultimately losing, causes. But the viability of missile defense does not come down to a life and death gamble on whether a Saddam Hussein, Mahmoud Ahmadinejad, or Kim Jong-il is rational or whether they will act responsibly if pushed to the wall. Instead, it hinges on the likely responses and actions of a large number of states, the probable effectiveness of defenses, and the abundant options available to US nuclear adversaries. Nor should a defense be assessed against a standard of "the possible," as in claims that a defense system will prove its value should it ever actually thwart a missile attack: should missile defenses "ever be used to defend a major American city, this investment will look modest indeed" (Daalder *et al.* 2000: 12). No option should ever be held to a standard of what could conceivably occur. The question for policymakers "is whether the insurance is worth the cost" (Wilkening 2000: 13). When all is said and done, any success-ful defense policy may only be as strong as the underwriting policies of deterrence, restraint, and engagement.

5

ON THE OFFENSIVE AGAINST TERRORISTS

Punishment and denial strategies

It is accepted axiomatically within the US policy community that deterrence principles do not apply directly to terrorists. This logic was articulated forcefully in the President's 2002 *National Security Strategy* statement when it asserted that "traditional concepts of deterrence will not work against a terrorist enemy whose avowed tactics are wanton destruction and the targeting of innocents; whose so-called soldiers seek martyrdom in death and whose most potent protection is statelessness" (Bush 2002a: 15). This assumption is not necessarily wrong. After all, terrorists live and thrive in the shadows of society without a fixed base of operations that governments can target; they do not control industries and cities that can be placed at risk through threats of punishment; they are by nature committed to changing the status quo and undermining deterrence; and they obtain legitimacy and derive strength only to the extent that they are successful in obtaining that objective. These are reasons, however, to believe that terrorists are not *easily* punished. In point of fact (as argued in Chapter 2), terrorists can be *punished* if only when they are *denied* their capabilities to conduct operations.

The strengths and limits of counter-terror strategies that rely upon punishment and denial principles can be obtained by exploring three interrelated perspectives that policymakers and scholars embrace to understand the terror problem. The first perspective assumes that terrorists are rational, coherent actors. It is invoked when analysts ask whether *punishment* is effective against terrorists given their predispositions and priorities. The second perspective assumes that terrorists are rational, coherent actors with state benefactors and popular constituencies. It is invoked when analysts ask whether terrorists are empowered or constrained, instead, by their reliance on states and societies. More specifically, it is invoked when analysts ask: (a) whether terrorist groups can be *denied* their objectives indirectly with acts of *punishment* aimed at state or public support of terrorism; or (b) whether acts of *denial* or *punishment* aimed at depriving these groups of their (state or public) support inadvertently boost it. The third perspective assumes that terrorists operate as organizations that are vertically and/or horizontally structured. It is invoked when analysts ask whether terrorists can be *denied* their objectives through the destruction or coercive disabling (denial or punishment) of specific parts of a terror organization or network.

From each of these perspectives an argument can be made that deterrence will and will not work against terror groups. But the second perspective—focused upon state and constituent support for terror groups—presents the greatest opportunities *and* risks for US offensive counter-terror strategies.

Perspective one: the rational, coherent actor

Even people who take their own lives are pursuing a rational course under some circumstances: a person suffering a terminal disease can rationally choose suicide to avoid the pain, financial cost, and anguish (personal and familial) of a prolonged decline to an inevitable death. Terrorists, too, can choose to kill themselves if that is required to serve some higher purpose (or obtain rewards in the next life). Thus, it is certainly unproblematic to accept the rationality of terrorist *leaders* who organize and plan suicide bombing campaigns that send *others* to their deaths. The sacrifice of these leaders is small and their gains from these attacks are possibly great.

When terror leaders are assumed to be rational, the *potential* effectiveness of counter-terror punishment and denial-based strategies becomes apparent. The effectiveness of these strategies depends on two factors: (a) whether terrorists respond to imposed costs; and (b) whether terrorist goals lend themselves to punishment.

Do terrorists respond to costs?

The effectiveness of a punishment strategy that directly targets terrorists hinges in part on whether states can make the punishment fit the crime. More precisely, the viability of the strategy depends on whether governments can impose retaliatory costs that exceed the benefits that terrorists obtain from an attack.

That terrorists respond to these costs is clear when these groups seek to avoid the consequences of their actions. For example, by operating with front names (as Fatah did when undertaking operations in the 1970s under the name of "Black September"), groups seek to escape the negative publicity and potential retaliation that a terror attack will bring. A sensitivity to imposed cost is also apparent when groups try to evade punishment simply by refusing credit for their operations. The US had to do a great deal of detective work to determine al-Qaeda's responsibility for the attacks on the US embassies in Africa and elsewhere. In fact, with considerable evidence that Osama bin Laden directed the World Trade Center attack, his public references to his role in the attack remained oblique. Generally speaking, the pattern since September 11 has been for groups not to acknowledge responsibility for attacks: terror groups claimed responsibility for eighty percent of terror attacks in 2001; by 2004, only thirty percent did so (Whitlock 2005a).[1] Further supporting the argument is evidence that terrorist leaders protect what they value. As the Taliban regime crumbled in Afghanistan in the wake of a US bombing campaign, neither the Taliban nor al-Qaeda leadership displayed their vaunted willingness to fight to the death, and seemed determined to postpone a final showdown with the West. Similarly, Palestinian militants targeted for death by Israel often went into hiding or surrendered to Palestinian authorities (David 2003: 143–4). Evidence of this cost sensitivity is found, too, when terrorists calibrate their violence to avoid losing public support or provoking an all-out government response. For instance, the Abu Musab al-Zarqawi organization (al-Qaeda in Iraq) backed away from the publicized beheading of Western captives in Iraq apparently in response to concerns that these actions were alienating the organization's support base. Moreover, the Palestinian Islamic Resistance Movement (Hamas) apparently rejected the use of chemical weapons against Israeli targets out of concern for weakening international support for the movement and provoking devastating Israeli military retaliation (Dolnik and Bhattacharjee 2002). It also moderated its behavior in times of weakness to placate the Palestinian Authority, maintain public support, and/or avoid confrontations with Israel (Mishal and Sela 2000).

Is it possible, however, for the US (or any other government) to manipulate the cost imposed on terrorists to deter them? In addressing that question, deterrence theorists stress the importance of crafting retaliation to deliver the right message. The costs imposed on the offending party should be quick (so that the target links cause and effect), severe (so that the costs to the target exceed the benefits), and consistent (so that the targets do not assume falsely that they can escape punishment). Only then, can an actor acquire the necessary reputation for resolve that can induce the target to abstain from action. This Cold War-era line of argumentation was adopted by those who supposed that the events of September 11 were encouraged by a feeble US response to terror strikes in the years leading up to the World Trade Center attack. To quote US Vice President Richard Cheney in this regard, "time and time again . . . the terrorists hit America and America did not hit back hard enough" (Coates 2005). The US did not respond directly to the attacks on the US military barracks that killed hundreds of marines in Lebanon in 1983, the Khobar Towers in Saudi Arabia in 1996, and the USS *Cole* in 2000. In fact, the US answer to the 1983 attack was to exit Lebanon. Assuming, then, that a failure to retaliate led terrorists to believe that they would not be held accountable for their actions, some have even argued that terrorist action should invite a hardening of a government's position on issues of concern to terrorists. For example, right-wing critics of Israel's Sharon government maintained that

Israel should have held its ground in Gaza (and not withdrawn in 2005), given the unwillingness of Palestinian groups to renounce terrorism, to avoid the appearance that Israel left under fire. Here, a lesson is ostensibly offered by President Kennedy's handling of the Cuban Missile Crisis. Regardless of the merits of the Soviet negotiating demand that the US remove its (outdated) missiles from Turkey in exchange for the Russian removal of missiles from Cuba, the Kennedy administration resisted any formal linkage between the two actions.

The case for retaliation is compelling in theory. In reality, efforts to build a reputation for resolve can fail because terrorists can read any number of meanings into an act of retaliation.

First, retaliatory actions can be interpreted as a victory for targeted groups when not resulting in their defeat. Given the overwhelming military superiority of the retaliating government—a reality that compels terrorists to resort to terror and avoid conventional military confrontations—any encounter between the mighty goliath and a sparsely armed, irregular force that fails to defeat it can be "spun" into a victory by the smaller force. In a sense, the latter is advantaged by low expectations. A dramatic example of this psychological dynamic is the 1968 battle of Karameh in the Jordan Valley. There, a relatively small Palestinian guerilla force, supported by Jordanian artillery, held off Israeli armored forces in intense hostilities that destroyed a number of Israeli tanks and killed dozens of Israeli soldiers. It forced, as legend has it, an Israeli defeat. In military terms, the battle was insignificant and the Palestinians suffered far more casualties than the guerillas inflicted in return. But it was unquestionably a political victory for Yasser Arafat and his Fatah organization. Coming as it did on the heels of the six-day rout of Arab military forces in the prior year, it was taken as an important sign that Palestinians could take matters into their own hands and accomplish what their state supporters could not. Arafat and Fatah capitalized on the victory to become the dominant force in the Palestine Liberation Organization (PLO).[2]

Second, retaliatory acts can be interpreted to suggest that the retaliating party lacks commitment and resolve, because of limits that are inherent in those acts. An act of retaliation is not, by nature, intended to defeat the adversary but to dissuade it, given the impossibility or cost ineffectiveness of a stronger response. Will the target group be more impressed, then, by the ferocity of the retaliatory response or its tepidness? The proverbial cup is "half empty," not "half full," when the recipient concludes that the retaliating country would have attacked with greater strength, *if* it could have. Thus, the 1998 launching of US cruise missiles on a pharmaceutical plant in the Sudan and empty training camps in Afghanistan, in retaliation for the US embassy bombings in Africa, could have sent the "wrong" message to al-Qaeda. Apart from the relatively paltry commitment of US resources that was involved—an effort that deliberately put no US military lives at risk—the attack could have signaled how little the US actually knew about al-Qaeda terror operations. In the first instance, the US struck a peripheral target with questionable links to the al-Qaeda network and, in the second, a target that was attacked too late to kill members of the organizational leadership that had assembled there. In point of fact, attacks that are intended to "send a message" are an implicit confession of weakness, that is, a signal that the sender cannot impose its will upon the target. The truth is that predominant parties do not worry about sending messages: they do what they want, when they want. That they do not send messages is the message.

Third, retaliatory actions by a government can be interpreted in light of its subsequent actions. The aggressiveness with which Israel responded to terror attacks and the growing ineffectiveness of terror attacks in the second intifada (al-Aqsa intifada or armed insurrection, 2000–5) did not prevent Hamas from claiming credit for forcing the Israeli withdrawal from Gaza and from benefitting electorally from a popular perception that the terror campaign led to an Israeli defeat. In the days leading up to the withdrawal, officials of Hamas gathered under its banner before the news media to pronounce their right to exist as a militarized force and to claim responsibility for the Israeli pullback (Associated Press 2005c; Wilson 2005a). With the withdrawal of Israeli forces, Hamas held rallies and military parades to celebrate the victory and its leaders appeared on television throughout the region to ensure that Hamas was properly credited for its accomplishment. Inasmuch as no acts of retaliation are strong enough, and no time interval is wide enough, to separate an act of terror from an eventual Israeli concession, what message will be sent, in the end, when Israel does concede ground? Whatever the military circumstances and despite the delays, the bottom line for Hamas is that Israel *did* withdraw.

Fourth, retaliatory attacks can be interpreted to suggest that the retaliating party will exhaust or defeat itself. Credibility—a belief that a party is willing and capable of supporting its commitments—is a depletable commodity. There is no guarantee that the credibility bolstered through retaliation will endure. The lesson that Saddam Hussein and Osama bin Laden learned apparently from the premature US exodus from Lebanon in the early 1980s and Somalia in the early 1990s is that the US lacked staying power (see Gordon and Trainor 2006: 66; Trager and Zagorcheva 2005: 104). For that matter, Saddam Hussein was sufficiently impressed by the scenes of US helicopters airlifting personnel from the roof of the US embassy in Saigon, when it fell to the Communists in 1975, that Iraqi television stations broadcast those images in the days before the 2003 invasion of Iraq (Laird 2005). This is ironic given that the US invested tremendous resources in efforts of marginal strategic importance to the US—in the case of Vietnam, considerable costs over a period of a decade in what was increasingly regarded as a failing effort. Could not a learned lesson of

Vietnam have been, instead, that the US would stick to its guns even when there appeared little hope of a satisfactory resolution, when the only thing keeping the US involved was its commitment to a commitment? For all the talk in US policy circles about the importance of maintaining US credibility, what counted in the end was less what the US did and more what US adversaries believed was true based on their peculiar views and interpretations of anecdotal evidence. For this reason, actions meant to strengthen US credibility can weaken it. Supporting this assessment are bin Laden's thoughts in a 2004 videotaped message: "All that we have to do is to send two mujahidin . . . to raise a piece of cloth on which is written al-Qaeda in order to make the generals race there to cause America to suffer human, economic, and political losses" (quoted in Mueller 2005: 40). A bold and aggressive US retaliatory effort can be viewed by terrorists as a useful US overreaction to provocation—a promising sign that the US will eventually defeat itself.

This is not meant to imply that the deterrence problem is a "signaling" problem. The effectiveness of deterrence does not hinge on whether messages are effectively conveyed to terror groups: in fact, no amount of retaliation might dissuade some groups from attacking their chosen targets. Indeed, recent terror assaults on Western targets have occurred though placing parts of the global terrorist infrastructure at risk. Those who attacked the London subway system were not swayed by the increased scrutiny and targeting of radical Islamic groups within Britain that such an attack would invite even though Britain was arguably a valuable base from which these groups could organize and rally the faithful (Coll and Glasser 2005a). Whether a punishment strategy works depends upon the goals that terrorists pursue and their commitment to these objectives.

Do their goals expose terrorists to punishment?

It is widely assumed that, by virtue of their goals, terrorists are immune to punishment. Certainly, there is merit to the argument that threats cannot deter those who pursue grand objectives and see benefits in martyrdom. After all, how can you punish someone with bullets and bombs whose path to glory is an explosive and fiery death? Isn't this like trying to punish a masochist with pain?

That these groups are susceptible to punishment is indicated, however, when terror groups pursue goals that governments can accommodate (Byman 2003; Cragin and Daly 2004; Trager and Zagorcheva 2005). In principle, it seems that governments can impose sufficient costs to deter the many terror groups around the world that pursue *limited* objectives and have relatively little to gain from attacking those governments or their interests. This is demonstrated by the actions of the secular, Liberation Tigers of Tamil Eelam (LTTE or Tamil Tigers)—a rebel group of the Tamil ethnic minority that seeks independence from the Sinhalese majority in modern-day Sri Lanka. The Tamil Tigers have almost always struck within Sri Lanka given fears within the organization that attacks outside the country might provoke retaliation against Tamils abroad who fund and equip the Tigers (Graeme *et al.* 2004: 197, 202–6). Indeed, the Tigers expanded their war to include Indian forces and targets only after India effectively became a party to the conflict by supplying peacekeeping forces to Sri Lanka (Crocker *et al.* 2004: 131). But it is not just the costs of retaliation that dissuade the Tigers from expanding the battle front. The focus of the LTTE upon the interests of the Tamil minority within Sri Lanka reduces the value of striking targets abroad. Because the LTTE's main objective is bringing direct and indirect pressure to bear on the Sri Lankan government to alter its policies, strikes outside of Sri Lanka offer lesser benefits for the LTTE than strikes within Sri Lanka. The same logic induced Hamas to focus attacks on Israel and to avoid strikes on non-Israeli targets. Whereas Hamas had sometimes threatened the US for its complicity with Israel,[3] Hamas had opportunities to mobilize the Palestinian population against the US and to strike US targets outside the Israel–Palestine theater if that were its goal. That Hamas demurred despite provocations is telling. The restraint shown by Hamas reflects its belief that the costs of expanding the conflict outweigh the benefits *given the low priority* of US targets.[4]

Retaliatory deterrence is facilitated, as well, when governments are positioned to make concessions to terrorists to satisfy their *intermediate* objectives. This is illustrated by Israel's relations with Hamas. From Hamas' standpoint, the Israeli occupation does not reduce to Israel's post-1967 positions in the West Bank and Gaza: Israel, by definition, is an occupying state and Zionism—the founding principles of the State of Israel—is the ideology of occupation. Thus far, no Hamas leader has been willing publicly to renounce the use of violence against Israeli civilians or concede the ultimate goal of establishing an Islamic state in all of Palestine (Erlanger 2006). But Hamas is willing to pursue goals short of this ambitious objective and, within the broad constraints of the organization's ideology, has adapted to political circumstances and the realities of power.[5] Consequently, when Hamas agreed in 2005 to a ceasefire with Israel to facilitate an Israeli withdrawal from Gaza, Hamas was vulnerable to retaliation from Israel. Because Hamas had positioned itself to grab credit domestically for bringing occupied territory under Palestinian control and to gain militarily from a loosening of Israel's grip over Gaza and border crossings, Hamas was susceptible to punishment by Israel if it were to slow or halt its withdrawal. In principle, punishment was possible inasmuch as Israel was conditionally willing to concede something that Hamas valued and Hamas had pinned its credibility, to some extent, on obtaining what Israel then possessed. (If Hamas was also seeking a ceasefire to forestall further Israeli military action in Gaza, it was clearly responding to punishment.)[6] Punishment could become more effective against Hamas now that it has assumed formal governance of the Palestinian territories (as the winner over Fatah in the 2006 legislative elections). Hamas might seek to avoid violent conflict and perhaps cooperate to some degree with Israel and rival Palestinian groups (and Israel via rival Palestinian groups) if held accountable—through a loss in electoral support or legislative coalition partners—for a lack of economic opportunity, drops in external revenue and aid, increases in violence and chaos in Palestinian territories, and a failure to obtain Israeli territorial concessions or prisoner releases.[7]

This does not mean that a basis for accommodation exists between the US and all terror groups, including al-Qaeda. The US is unlikely (if not unable) to pursue a foreign policy that will placate the al-Qaeda leadership even in the short term. There appears little room for compromise between the US and al-Qaeda given its absolute demands—its unwillingness to accept a Western presence in the Middle East, its insistence that Israel exists as an arm of Western imperialism, its assumption that Western-leaning governments within the region are held in place by Western support, its interpretation of Islam that requires a rejection of Western practices, its strongly held belief that Islam is under siege and fundamentally threatened by hostile cultural forces, and, most importantly, its belief that the US is the "serpent's head" that is "the root of every evil in the world" and that the annihilation of the US will facilitate the downfall of anti-Islamic forces everywhere (Schweitzer and Shay 2003: 25). In turn, the US unwillingness to concede control of the Middle East and its oil to fundamentalists who reject the West on principle and view Israel as an illegitimate entity means that the US and al-Qaeda are parties to a zero-sum contest. As Byman (2003: 147) puts it:

> because of the scope of its grievances, its broader agenda of recti-
> fying humiliation, and a poisoned worldview that glorifies jihad as
> a solution, appeasing al-Qaeda is difficult in theory and impossible
> in practice. It is hard to imagine what would suffice, as so many
> U.S. interests are involved that even significant policy changes
> would be only the tip of the iceberg.

As he recognizes, then, any effort by the US to placate al-Qaeda would be viewed by it as a tactic meant to foster nefarious US goals: for example, promoting Middle East peace talks "would be interpreted as an attempt to force docile Muslim regimes to legitimate Zionist imperialism rather than as a step forward in the Palestinian cause." In dealing with al-Qaeda, the problem for the US is what it "is," not what it "does."

This view that there is little to no basis of accommodation between Islamic radicals and the West (that could be harnessed for coercive leverage) is not universally held. In a provocative, well-documented, and much-publicized study (growing out of a research project at the University of Chicago), Pape (2005) argues that the logic of "suicidal terrorism" is not fundamentally a religious/ideological one. Instead, he argues it is a "strategic" one in the sense that suicide terrorism is intended to accomplish the tangible, ostensibly limited purpose of ending an "occupation"—ridding Saudi Arabia of US occupation, Iraq of US occupation, and Palestine of Israeli occupation. Bolstering his argument is that fundamentalist Islamic groups have generally focused their efforts on immediate, national goals— overthrowing venal, secular governance in their own countries. Even groups such as Hamas and Islamic Jihad concentrate their appeals on Israeli occupation. In point of fact, suicide terrorism, as a modern-day phenom- enon, emerged in the wake of Israel's 1982 occupation of Lebanon with a Hezbollah truck-bomb attack that killed 241 US marines in Beirut; and, more generally, a growing US presence in the Middle East has coincided with escalating attacks on the US and allied countries. Moreover, evidence shows that suicide terrorism is not strictly an Islamic phenomenon. Committed practitioners include the Tamil Tigers who, commencing in 1987, waged a prolific suicide bombing campaign against a wide variety of civilian government, military, business, and religious targets that most notably killed the Indian Prime Minister and the President of Sri Lanka in the early 1990s.

Still, Pape's thesis is problematic (indeed, nonfalsifiable) in large part because of his fuzzy definition of key terms. The first of these problematic terms is "strategic." It is one thing to say that radicals are strategic in that the objectives of these groups are rationally served by their strategies; it is quite another to suggest that "modest" objectives are pursued by these groups that belie their harsh ideological demands and claims. Although various groups—including the Tamil Tigers—have used suicide terrorism to accomplish their objectives, al-Qaeda's worldwide terror campaign actu- ally draws deep inspiration from ideological and religious goals. How else to explain the global reach of these campaigns and their appeal; bin Laden's willingness and ability to surround himself with individuals of different races, ethnic backgrounds, and nationalities (including Ayman al-Zawahiri, bin Laden's Egyptian top lieutenant); and the extent to which participants view linkages between the struggles in far-flung places, from Palestine, Jordan, Iraq, Saudi Arabia, Egypt, and Afghanistan to the Balkans and Indonesia? The suicide bombing of the London subway, with an apparent al-Qaeda link, is a dramatic case in point: British citizens of Pakistani descent attacked civilian targets in London ostensibly in retaliation for British support for a US-led operation in Iraq. Furthermore, contrary to Pape's claim that al-Qaeda has primarily Saudi-centered origins and griev- ances, al-Qaeda has not focused its actions exclusively on the Gulf and has been involved extensively in conflict in other parts of the world, including the Israel–Palestine conflict. This is indicated, for instance, by the al-Qaeda operation aimed at Israeli tourists in Mombasa, Kenya (belying Pape's claim that al-Qaeda's suicide bombers have never attacked Israel).[8]

Inevitably, too, the question is, "what makes these, 'occupations'?" The US is hardly the inspiration behind the Saudi style of governance—monarchical, hierarchic, and theological—or much of Saudi Arabia's domestic and foreign policy. The US enjoys basing rights in Saudi Arabia, but these forces do not sustain the Saudi regime in power and remain in Saudi Arabia at the discretion of its leadership. Indeed, Pape considers terrorist violence against the US throughout the Middle East—in countries such as Turkey, Kuwait, Qatar, Bahrain, the United Arab Emirates, and Oman in which the US military presence is modest at best—to be reactions to US occupation. For that matter, he does so despite offering an explicit and demanding definition of foreign occupation (2005: 46) that centers on objective determinants ("a foreign power has the ability to control the local government independent of the wishes of the local community") *and* subjective determinants (a recognition by the effected population that such control exists).[9] In practice, his designation of a country as occupied is based on *his* judgment that terrorists *believe* their country is controlled by another. It is ironic that Pape's conclusions hinge on subjective beliefs of effected groups, then, when he is unwilling to accept at face value the professions of implacable hostility toward the US articulated consistently by al-Qaeda leaders. If Pape had built on his analysis of subjective beliefs, he would have better appreciated the ideological roots of the current conflict.

Ultimately, more questions are raised than answered by his analysis. Why is it that the intersection between Islam and the West is viewed by al-Qaeda in imperial rather than conjunctive terms, as an encroachment rather than an interrelationship? The US has maintained troops in a wide variety of countries, often over long periods of time (e.g. South Korea); it occupied Germany and Japan after a brutal war that was ultimately meant to exact a high civilian toll. Indeed, sixty years after the end of World War II, the US still maintains military forces in Germany and Japan and plays a critical role in shaping the security policies of both countries. Why are these relationships not currently considered occupations? It is easy to view the rejection of a Western presence in Saudi Arabia or a US alliance with Egypt by Islamic extremists as part of a deeper grievance against corrupt and blasphemous regional governance linked inextricably to relations with the West. But it is hard to view these relationships as occupations. The US actions that Pape sees as motivating suicide bombing are hardly exceptional to the Middle East or a particular moment in history. What is exceptional is the suicidal reaction to them.

The link between "occupation" and terrorism must also be questioned because Pape's analysis rests on the unsupported assumption that suicide terrorism is a single and unique form of terrorism.[10] Pape might have reached different conclusions had he recognized the dissimilarities among groups that practice suicide terrorism (Atran 2006: 131) or assessed suicide attacks that were meant *primarily* to kill civilians (rather than to destroy government or military installations, frequent targets of the Tamil Tigers), *failed* suicide bomb attacks, attacks that involve high-risks to the attacker and are *effectively* suicide attacks, or, for that matter, *all* terror attacks. Illustrating that al-Qaeda was not purely focused on American occupation of the Gulf and *did* have its eye on Israeli targets, for instance, are the travels to Israel of Richard Reed, the notorious shoe-bomber who attempted to blow up an American Airlines flight between Paris and Miami in 2002 and, more recently, the arrest by Turkish authorities of ten people, in August 2005, with suspected links to al-Qaeda in a plot to bomb Israeli cruise ships along the southern Turkish coast (Vick 2006). The last of these incidents is too recent to be included in Pape's analysis but, even if not, it would have been excluded. This reflects a systematic bias in the data collection. Certainly, one of the reasons that al-Qaeda has not been more involved in suicide attacks on the Palestine front is the difficulty in gaining access to Israeli targets. Thus, failed terror attacks of all sorts might provide a better indicator of al-Qaeda's *intent* than actual suicide attacks.

The link between occupation and terrorism further erodes when the data are examined in relative rather than absolute terms. Although Pape draws his conclusions based on overall numbers of suicide terrorist incidents, different conclusions are supported when the question becomes whether Islamic groups had a greater *tendency* to employ these tactics when compared to other groups.[11] Based on his data, the answer is clearly, "yes." In the 1980–2003 period, thirty-three percent of Islamic groups opposed to a democratic "occupation" resorted to suicide terrorism compared to five percent of non-Islamic groups. For that matter, when controlling for Pape's preferred explanation for acts of suicidal terror—a religious "difference" between an indigenous population and an alleged democratic occupier— by only examining conflicts in which these contenders differed in religion, Islamic challengers are still more likely to resort to suicide terrorism than other groups are: under these circumstances, forty-three percent of Islamic groups resorted to suicide terrorism compared to fifteen percent of non-Islamic groups.[12]

Even relying upon Pape's inferential approach (and counting suicidal terrorist incidents) suggests that Islamic groups are behind most of the rapidly growing number of acts of suicide terrorism worldwide. Post-September 11 attacks account for over eighty percent of all suicide bombings, with attacks in Iraq largely responsible. With hundreds of suicide bombing attacks producing thousands of casualties in Iraq since the initial bombing of the Jordanian embassy and UN compounds in Baghdad in August 2003, the number of suicide bombings between 2004 and 2005 exceeded the numbers of attacks in Pape's entire (1980–2003) sample (Atran 2006: 127, 133). By mid-2005, the *monthly* numbers of Iraqi bombings had reached the *total levels* incurred in the Israel–Palestine conflict in the previous twelve years (Eggen and Wilson 2005). The impact of these numbers is not lessened by the realization that non-Islamic groups have also resorted to suicide terrorism. Indeed, with the recent reduction in suicide terrorist acts committed by the LTTE—Pape's notable exception— suicide terrorism became virtually the exclusive prerogative of groups with an Islamic composition.

This is not to say that Islam itself is the source of terrorism; instead, it is meant to suggest that bin Laden and other extremists can use Islam to promote a compelling message that plays off national-level grievances (perhaps, perceptions of "occupation"). Such interpretations of Islam provide a powerful motivation for believers and a tool for capitalizing on nationalist, social, and economic distress of Muslims around the world. Islam elevates and legitimizes these grievances by giving them a deeper, global significance: "the historical narrative, however stilted or fictitious, translates personal and local ties within and across small groups into a profound connection with the wider Muslim community" (Atran 2006: 136). As a result, Islamic insurgents in faraway places such as Thailand see the images of Iraqi insurgents and are motivated to add forces and new military fronts for the cause (Nakashima 2005). After all, the Palestinian cause is embraced throughout the Middle East because the conflict between Israel and Palestine serves as a touchstone in a broader pan-Arab and/or pan-Islamic struggle. That the issue resonates so widely can be attributed to the deprivation of life, liberty, and property of the Palestinian population by the Israeli government yet social, political, and economic abuses have been a fact of life throughout the Middle East. Why is the abuse of Palestinians viewed (by implication) as more repugnant than the suffering of people in the Darfur region of the Sudan, the mass killing of Kurds under Hussein in Iraq, and so forth? Rhetorical support for the Palestinian cause shields regional leaders certainly from having to contend with abuses and inequities that stem from these leader's own oppressive and nonegalitarian practices. Still, the fact that people respond to the issue and that leaders can credibly raise the concern speaks to the broader implications of the Palestinian cause and the extent to which a national cause can reinforce a regional one (and vice versa). Likewise, US intervention in Iraq provides a cause around which Iraqis, but also Egyptians, Saudis, and Palestinians, can rally.[13] To say, then, that each group is only concerned about its specific battlefield is to misunderstand the nature of the problem that the US and its allies face worldwide.

In an important way, then, the US position relative to al-Qaeda is similar to Israel's position relative to Hamas. Support for Hamas is unquestionably linked to popular disdain for Israel's retaliatory policies in the West Bank and Gaza, military roadblocks that disrupt commerce and movement within those territories, and Jewish settlements that claim Palestinian lands.

By playing on these grievances, Hamas has been able to position itself in the vanguard of the effort to end the Israeli occupation. But Hamas' nationalist objectives are linked to a broader Islamic commitment, as expressed in its rhetoric and uncompromising political stances that include an unwillingness to negotiate directly with Israel or concede its right to exist. Indeed, within the charter principles of Hamas, "the territorial objective" is "defined as integral to Islamic duties and beliefs" and "giving up any part of Palestine" is viewed by Hamas as tantamount to "giving up part of its religion" (Mishal and Sela 2000: 44, 183). Despite the danger in reading too much into an official document, these words do mean something: they were selected intentionally over others. To focus only on Hamas' minimalist (immediate) objectives, then, is to misunderstand the organization (as currently constructed) and understate its willingness to pursue its maximalist objectives as circumstances require or permit. By capitalizing, for instance, on economic distress and popular unrest in the territories or in seeking to divert public attention from the organization's failures at governance, Hamas (or one of its wings or offshoots) could push the argument, at some point in the future, that an answer to the problems of Palestinian society lie only within violence, a religious transformation, and a broader territorial solution.

In the end, the US must try not to exacerbate its problems through actions that inflame passions, inspire resistance, and bring new recruits to the cause, but it will not eradicate a threat from terror groups that cannot be accommodated with realizable adjustments in US policies. Thus, it is hard to be optimistic that a retaliatory strategy (aimed at terrorist assets) will succeed when intended to shape the calculations of al-Qaeda leaders. For al-Qaeda to hold its fire and seek some form of compromise with the West requires that the group renounce its objectives, change its identity, and accept defeat. That al-Qaeda could be *compelled* to do so through fear of punishment seems unlikely.

Perspective two: the political context

A better understanding of the rational terrorist can be achieved by studying its political environment. Terrorist groups operate within territory that is nominally under the control of states; often acquire resources, logistic support, and training from states (Byman 2005: 59–66); and draw recruits and material from a supportive population. Whether these groups maintain significant leverage when dealing with states and public supporters or whether the reverse is true is often a matter of dispute. The collective opinion on the subject suggests, then, that the political context promises both the greatest leverage and risks to outside governments seeking to use deterrence instruments to control terrorist behavior.

State support

When targeting state sponsors of terrorism—whether active promoters or reluctant government hosts of a terror group—a punishment-based strategy is meant to deny terrorists access to critical resources or territory from which to operate. Even threats or attacks on a terror group's assets can be employed for that purpose, with the intent to send the message that future attacks will directly target the supporting government should it not end its active, tacit, or involuntary aid to the group. Therefore, when directed at government *or* terrorist assets within a host country, retaliation could be meant to signal that the host government: (a) is directly responsible for the actions of a terror group; (b) is ultimately liable for terror actions committed from its territory; (c) is better equipped than the retaliating government to disable the terror group (by merit of a shared culture, language, and/or ethnicity, access to terror bases, and local intelligence); and/or (d) must accept the costs of defeating the terror group. The message could be underscored were retaliatory acts to impose reputation costs (at home and abroad) on the targeted government by openly challenging its ability to protect its citizens, pursue its policies, and safeguard sovereign national territory. The sender hopes obviously that the targeted government will be induced to curtail its ties to the terror group.

That a punishment strategy can work against state sponsors of terrorism is apparent from evidence that economic sanctions forced Libya to distance itself from terror operations. In the 1990s, Libya sought to head off a tightening series of UN sanctions adopted against Libya for its failure to renounce, and make amends for, its terrorist past. The UN demanded that Libya accept responsibility for the bombing of a US Pan American flight over Lockerbie, Scotland, in late 1988, cooperate with the investigation into Libya's involvement, compensate the attack victims, turn over Libyans accused in the bombing, and end its support for terror groups. A country that, in 1989, was said by the US government to be supporting over thirty terrorist networks and revolutionary governments worldwide was said by the US State Department in 2001 to have "curtailed its support for international terrorism" (O'Sullivan 2003: 208). That a punishment strategy can succeed is also found in evidence that US threats induced the Sudan and Yemen to adopt less accommodating policies toward terror groups (leading to bin Laden's expulsion from Sudan in 1996 and Yemen's cooperation with the US in targeting local terror units).

Perhaps the most oft-cited case in support of a punishment strategy against state support for terrorism is the ouster of Palestinian groups from Jordan in 1970–1 (the Jordanian civil war). These groups were a strong and visible presence in Jordan, where they operated as a virtual state within a state, flaunting their power in public displays—landing and blowing up three hijacked airliners in Jordan in the incident that sparked the civil war.

Despite the strong Palestinian presence in the country and King Hussein's paralyzing fear that moving against Palestinian militants would regionally isolate Jordan and doom its monarchy, Jordan violently confronted these groups. Hussein's position had become untenable in part because Israeli retaliation against Palestinian targets in Jordan had forced the King to take sides in the conflict or lose control of his kingdom. Hussein acted against these groups, it appears, when the costs of inaction became prohibitive (Astorino-Courtois 1998). Around three decades later, Yasser Arafat, as head of the Palestinian Authority, was ironically placed in a similar position when various radical groups staged attacks on Israeli targets from the West Bank and Gaza. Although Arafat insisted that he was incapable of controlling the militants (given provocative Israeli actions and the weakened position of his security forces), he was able to get militants to observe a ceasefire after September 11 when he feared that ongoing acts of terror would hurt the Palestinian cause by associating it implicitly with al-Qaeda (Bloom 2005: 21–2). Similarly, his successor, Mahmoud Abbas, was able to exert some pressure on extremist Palestinian factions to halt their attacks on Israeli targets. He clearly feared the threat to his leadership should Israeli retaliation destabilize conditions in the occupied territories and further delay the march toward statehood and the consolidation of Palestinian rule.

Even Afghanistan under the Taliban offers an example that a punishment strategy aimed at state sponsors of terrorism can work. It appears that Taliban leaders wanted to limit the scale and location of al-Qaeda attacks on US interests because of concerns about US retaliation. There were serious debates among these leaders over the wisdom of so closely affiliating the Afghani cause with al-Qaeda's global ideological and terror campaigns (apart from the resentment felt by the Taliban rank and file toward the Arab al-Qaeda presence in the country). Thus, the US attack on Afghanistan must serve as a warning to governments that employ surrogates for actions that are dangerously provocative when taken directly by states.

But a punishment-based strategy against state sponsors has significant liabilities. For a number of reasons, it is hard to punish states for their alleged support of terrorist activities.

First, effectively retaliating against state sponsors of terrorism is complicated by the considerable problems of determining indirect responsibility for an attack (within the time period in which retaliation remains useful). Placing blame on a particular state is difficult when competing groups take credit for an attack, when no one group—of those backed by different states—takes credit for an attack, and when groups pool their resources and coordinate their efforts so that "leads" point in multiple directions. Byman (2005: 305) speaks of the "Talmudic debate" within the Clinton administration as officials weighed pieces of evidence to determine degrees of governmental responsibility for an attack. Consequently, a state can hide behind a veil of ignorance and deny that terrorist organizations operate from the state's territory, or else the state can claim that it is doing its best, given limited resources and pressing demands, to combat terrorism despite a limited record of success. Who the instigators of a terrorist attack are and who are bystanders is often a matter of perspective. Of note, here, is that the Bush administration was far more impressed with evidence of Saddam Hussein's complicity in the September 11 attacks—despite weak evidence of a link and the adversarial relationship between Iraq and al-Qaeda's leadership—than evidence that support for al-Qaeda permeated the Saudi Kingdom and reached into its government.

Illustrating the role of perspective is the issue of Yasser Arafat's complicity in terror attacks on Israel from territories under the nominal control of the Palestinian Authority during the second intifada. The now widely accepted view that Yasser Arafat was tacitly if not actively supporting these operations, was resisted at the time by most Western governments. Arafat's culpability was denied despite evidence that secular Palestinian groups and, in particular, the al-Aqsa Martyrs' Brigades, the armed wing of Arafat's political movement, Fatah, participated in the attacks and that the Palestinian Authority was involved in weapons smuggling into the occupied territories. The latter evidence includes the seizure in 2002 of the Karine A, a freighter owned by the Palestinian Authority, loaded with prohibited weapons including rockets, mortars, and anti-tank missiles; it also includes documents establishing financial and personnel links between the Authority and groups engaged in terror attacks. Israel maintained that Arafat was duplicitous—he was insisting that he was doing everything in his power to reign in radical groups and abide by the terms of the Oslo agreement (under which the PLO renounced the use of force against Israel) while also supporting, funding, and actively directing terrorist operations against Israel. But the prevailing assumption was that Arafat was caught in the middle—that Arafat wished to move against radical groups—if only to consolidate his power—but lacked the political and military capabilities to do so. Although Arafat's government was not a "state," as strictly understood, his relationship with radical groups clearly demonstrates the problem in determining governmental responsibility for terror acts. From the standpoint of many Western states, Israel failed to produce the "smoking gun" to link Arafat directly to terrorist activities. For that matter, whether or not Arafat condoned and encouraged the creation and activities of terror groups to place heat on Israel, it was also true that groups such as the al-Aqsa Martyrs' Brigades had their own agendas and were loosely formed, without a centralized leadership that could order an end to the violence (Cordesman 2005: 185).

Second, retaliating against countries is potentially ineffective when terror groups can move among nations and/or play governments off against one another. Few worked harder at this strategy than Yasser Arafat (albeit with mixed success). He was able to maintain a formidable presence in Jordan through 1970–1 by relying upon his regional supporters to restrain Jordan from moving against him. Indeed, when Jordan finally did confront Palestinian groups, Syria intervened on their behalf. With a subsequent shift in the PLO's center of operations from Jordan to Lebanon, Arafat outmaneuvered the Syrians in the 1970s, during the Lebanese civil war, by establishing common cause with Lebanese leftists. In his decades of political maneuvering, Arafat survived and prospered though alienating, at one point in time, each of his important regional allies—Egypt, Jordan, Syria, Iraq, and Saudi Arabia. Similarly, the spread of the Kurdish population across sovereignties—especially, Turkey, Syria, Iraq, and Iran—made it difficult for any one state to control the activities of Kurdish groups. The Kurdistan Workers' Party (PKK), in engaging in a war against the Turkish government that took the lives of tens of thousands of people in the 1980s and 1990s, survived against an overwhelmingly superior military force in part by using sanctuaries in Iraq and Syria. Likewise, the ability to move across national boundaries has been critical to bin Laden who left the Sudan for Afghanistan in the 1990s and Afghanistan for the border areas of Pakistan (at least, so it seems) with the overthrow of the Taliban.

Third, retaliating against host governments is problematic if they are weak politically or militarily relative to terror groups and their public supporters within a given part of the country. Retaliating (directly or indirectly) against these governments might have no effect when terrorists operate from national territories that are only nominally under state control—refugee camps (Byman 2005: 66) or so-called black spots or lawless regions, as in the southern Philippines, parts of Africa, the tri-border area among Argentina, Bolivia, and Brazil, or the tribal regions of western Pakistan (Blum *et al.* 2005: 135; Takeyh and Gvosdev 2002). Retaliation against host governments might also be unproductive when they are politically disadvantaged relative to radical groups, and choose not to confront them. Because bin Laden and his allies benefit in western Pakistan from considerable tribal and radical Islamic support, the Pakistan government has been reticent to move—independently or in coordination with US forces—against al-Qaeda and Taliban elements located there (VandeHei and Graham 2006). Saudi Arabia was also reluctant to act when pressed by the US to quash domestic radicalism after the September 11 attacks. Given the support enjoyed by radical Islamic groups within parts of Saudi society and the legitimacy that the Saudi government acquired from adhering to fundamentalist Islamic practice, the Saudi government was extremely hesitant to take on these groups. The Saudi government allowed radical Islamic groups to recruit and raise money on Saudi territory, and adopted a somewhat benign attitude toward bin Laden's international exploits, as long as Islamic terrorist activities were directed at foreign targets. The Saudi position changed dramatically only when the government was compelled to redefine its interests when Islamic elements turned their weapons on Saudi Arabia in 2003 (Byman 2005: 223–38). Governments are similarly challenged when terror groups rely on nonreligious mechanisms to build a popular following. Groups such as Hamas in Gaza and Hezbollah among the Shiite population of Lebanon enjoy considerable popular support for their ardent promotion of "constituent" interests and their vast social support networks of schools, hospitals, and so forth. Retaliation could actually make matters worse, then, if reducing the targeted government's capabilities to challenge terror groups without increasing the government's willingness to act against them. For example, in decimating the security forces of the Palestinian Authority for their complicity with militant groups during the armed intifada, Israel arguably foreclosed the possibility that the Palestinian Authority would rein in these groups (Cordesman 2005: 168–71). The situation could worsen, too, if host governments respond to pressure and confront the terror groups. In the early 1970s, Israeli military pressure on the Lebanese government did induce it to move against PLO forces, but not without sparking a civil war that led to Syrian military intervention, and undercutting an agreement that had imposed some restraints on PLO conduct in and from Lebanon (Sayigh 1998). In 2006, the Israeli military offensive against Hezbollah in Lebanon raised serious concerns, once again, about whether the Lebanese government would be fatally weakened if it capitulated to outside demands and tried to disarm the militia or restrict its operations.

A party considering retaliation must recognize, then, that a weak host government might still be better than no government, or a likely replacement government, in territories from which terrorists operate. The possibility that things could be worse haunted Israel when it confronted the Palestinian Authority. Israel's perception that Arafat was weak and could not stop terror attacks arguably "saved the peace process in the early 1990s" (Kydd and Walter 2002: 276). The same possibility challenged Israel in its dealings with Arafat's successor, Mahmoud Abbas, when he insisted that his hands were tied and that he was unwilling to risk a civil war by disarming radical groups. For Israel, then, the challenge was to determine whether the Palestinian Authority was doing what it could to end terrorist attacks, whether Israeli forces could do a better job, and whether any gains from Israel's intervention would be offset by the further weakening of the Abbas government, or its collapse.

Fourth, retaliating might not work against governments that are effectively or ideologically bound to a terror group. When terror groups and the government exist in a symbiotic relationship, it is impossible to separate the interests of the state from the interests of the terror group to make coercive threats work. Inasmuch as the Taliban and al-Qaeda had become indistinguishable (interwoven financially, militarily, and politically), the US failed in its post-September 11 efforts to get the Taliban to turn over the al-Qaeda leadership to the US. Retaliating effectively against governments is at least as difficult when their links to terror groups are ideational in form. Because the fundamentalist leadership of Iran obtains legitimacy from its support for groups such as Hezbollah, leadership ties to these groups cannot be broken without de-legitimizing the regime. Of course, ideological ties overlap with political, military, and economic linkages: the Iranian government recruited Hezbollah for missions that included assassinating government opponents in Europe and giving Iran a presence on the Arab–Israel front; and Syria used like-minded Palestinian groups to apply pressure on Israel, when using conventional military force would have sparked a war that Syria could not win (Byman 2005). Governments with overlapping linkages to a terror group are even less inclined to confront it. By the same token, some imperviousness to coercion can result when radicals enjoy backing only within parts of a government. US pressure on the Pakistani government to rein in extremist Islamic groups has been thwarted by their supporters within the Pakistani military (especially the Interservices Intelligence organization which helped bring the Taliban to power) and unease at various levels of Pakistan's government with pro-US or pro-Western policies that place Pakistan at odds with sentiment throughout the Islamic world. Similar strains are observed in Saudi Arabia from where government (and private) money has moved through "religious charities" to terrorist groups around the world. Even somewhat disingenuous support—such as Saddam Hussein's embrace of Palestinian suicide-bombing operations (through aid to the families of suicide bombers) —could limit the effectiveness of retaliatory policies. That Hussein felt pressure to support terrorism to bolster his domestic position and pursue his international agenda in the first place suggests that he had much at stake politically in the furtherance of these policies.

Fifth, military retaliation might be ineffective against states that pursue terrorism as a national security strategy and are inclined to respond to retaliatory attacks with retaliation of their own. That these states are undeterred by *possible* bloodletting—arising from retaliation for their retaliation—is demonstrated by Libya's response to US retaliation for Libya's role in the bombing of a Berlin nightclub that killed two US serviceman. With the US attack on military and government targets in Tripoli and Benghazi in 1986 (leaving a number of people dead, including Moammar Qaddafi's adopted daughter, and coming close to killing Qaddafi), Libya responded with a wave of terror attacks. These included placing the bomb on board the Pan American flight that exploded over Lockerbie, Scotland, and bombing a French flight over Niger which resulted, together, in hundreds of deaths. Whereas 91 people were killed in terror attacks linked to Libya in the five-year period before the US attack, 491 people were killed in attacks with a Libyan connection in the five years that followed (Collins 2003: 6; see also Schweitzer 2004).

Sixth, military retaliation against states might not send the intended message when the "meaning" of a retaliatory attack is left for the target to interpret. Given the multiple meanings inherent within a single coercive gesture (as discussed above), the target could take a signal meant to communicate resolve as a message of weakness or lack of commitment. There is more room for interpretation when retaliation targets terror groups. Should the attack be taken as an implicit threat to the host country? Or, does it suggest that the host government is "off the hook"—in other words, that the retaliating government does not believe that the host can police itself and that it will not be held responsible, then, for terror acts committed from its soil? Such mixed messages were arguably sent to the Palestinian Authority by Israeli attacks on Palestinian groups in the occupied territories. Of course, retaliation always risks confirming the targeted country's prior apprehensions about the motives and character of the retaliating state. Then, the retaliating state can expect, at best, that it will receive no *additional* hostility from the targeted government after its territory has been struck.

Seventh, retaliating against states that support terrorists is made difficult because these states carry a negative burden of proof. To be effective, punishment strategies require that the punishment end once the target has complied. But how exactly does a government establish that it did *not* support a terror group, especially if the group were to continue to engage in attacks and outsiders have every reason to doubt the sincerity of the government's denials of responsibility? A government can only "prove" its innocence with a politically unacceptable gesture such as active cooperation in the US war on terror. If lesser concessions are invisible and the state's complicity in terrorist operations is assumed by default, a state might have little to gain by ending its support for a terror group.

Taken together, the evidence is that governments have had some success holding other governments accountable for supporting terror groups. That Libya was pressured to get out of the terror game is perhaps the most valuable case in point, though supporting evidence is found, as well, when governments offer "less-than-complete" support to terror groups. Certainly, deterrence is in play when the Iranian government supports these groups only in secret, within limits (e.g. when Iran deprives Hezbollah of certain weapons), to a declining degree (e.g. when Iran reduced its support for Shiite militants in the Gulf after the Khobar Towers incident allegedly to avoid US retaliation; Byman 2005: 261), or in local instead of global operations (e.g. when Iran encourages Hezbollah to act against Israel but not the US). Thus, retaliatory policies aimed at states offer hope to governments seeking to control the behavior and constrain the operations of terror groups located abroad. But the advantages of a punishment strategy are easily overstated. It is difficult to establish that an alleged state sponsor of terrorism acted with malice and that its punishment fits the crime; and then, coercion can fail when messages are left for the target to interpret and it is loath to capitulate. Coercion can fail, too, if efforts to deter terrorist attacks by targeting their government sponsors or hosts are effectively efforts to *compel* those governments to change their policies, when doing so requires important and/or visible political, economic, military, or ideological concessions. It is one thing for the targeted government to decide not to subsidize, train, or provide logistic support for a terror attack; it is another for the government to withdraw support for a terror group; and it is quite another for the government to take on a terror group and try to put it out of business. Ironically, governments that successfully coerce others to move against terror groups lose in the end if the pressured governments are weakened politically or militarily by the coercive effort, or it results in a failed or counterproductive anti-terror campaign.

For that reason, retaliating states can opt for a denial over a punishment strategy, as the US did, after September 11, when it overthrew the Taliban government of Afghanistan. But that experience also offers lessons on the enduring challenges of regime change (as discussed generally in Chapter 3). In Afghanistan, these challenges include bringing about national reconciliation, balancing diverse societal interests, building governmental institutions, and extending the government's control beyond the capital. Consequently, the Afghan experience suggests that this option is best regarded as a last resort.

Public support

Public support is critical to the terror enterprise. Because terrorists operate from a position of military weakness, they must pursue a *political* strategy that relies upon demonstration, coercion, and incitement (Hoffman and McCormick 2004: 247; Pape 2003: 345). Violence is employed by the terrorist, then, to elevate its standing, publicize a grievance, intimidate opponents, destroy public morale, disrupt an economy or society, produce government concessions (e.g. policy changes or prisoner releases), and/or provoke a government to over-react (that is, to overextend itself or adopt heavy-handed policies that validate the terrorist's message and reduce support for government policies). Given these objectives, terrorists must consider how their actions, or government retaliation for these actions, will affect the position of the terror group relative to society.

Recognizing the importance to terrorists of their popular base, governments can seek to deny terrorists access to key societal resources. They can do so by threatening those who support or might support terror groups or by targeting the portions of the terrorist infrastructure that are co-located with or tied to a civilian population. These include businesses or individuals that finance terrorist operations, safe houses, arms merchants, bomb-making factories in auto shops, educational or religious centers that are used for recruitment or indoctrination, and private residences. In seeking to deprive terrorists of these resources, governments might differentiate between terrorist and non-terrorist assets (and combatants and noncombatants) and employ force within established limits to avoid collateral damage. In this, governments are likely aware that even limited actions have deterrent consequences. Note, for instance, Israel's policy of destroying the homes of families of suicide bombers to dissuade others from becoming suicide bombers or supporting them. Still, governments might try to magnify these effects through indiscriminate action or liberal applications of force erring on the side of inclusion—that is, judging people guilty by association with terrorists or their financial enterprises. A possible benefit of indiscriminate force is that it is sure to inflict at least some direct or indirect costs on those who actually participate in terrorism.

A conventional wisdom, however, on the use of offensive (denial and punishment) strategies against terrorists is that battling terror groups on their home turf will fail. Terrorist groups can build a societal support base, that includes even those who are not otherwise predisposed to accept the group's message, by playing on their ethnic or religious ties to a population, a popular economic, social, or political grievance, the absence of an effective and moderate opposition to government policies, and/or their own good works within the community. This has been true of various Palestinian groups; and it is true of the Taliban, which is tied via its Pashtun ethnicity to parts of the Afghani and Pakistani population. In consequence, terror groups are positioned to hide their activities within the normal affairs of the community, to suffer the problems of the community, and to convince the community that the group can solve those problems.

Thus, attacking terrorist assets is said to be counterproductive. Palestinian terrorism in the occupied territories follows in a tradition—the Irgun in Palestine in the 1940s, the Algerian resistance in the 1950s, and the Uruguayan Tupamaro Liberation Front of the late 1960s and early 1970s—of using violence to provoke the enemy to expose it for "what it is." Frequently noted is that, by provoking retaliation, the terrorist can drive a wedge between a government and its people. Terrorists benefit when innocent individuals are picked up in dragnets and subjected to harsh interrogation, imprisonment, and arbitrary punishment, when fields are burned and property confiscated under suspicion that it is being used to support terrorist operations, when houses are destroyed because they have been used as platforms in attacks, when people live in constant fear of arrest or death, or when people simply happen to be at the wrong place at the wrong time and are caught up in the violence. The consequence, it is argued, is increasing receptivity to the terrorist message—that the adversary is brutal and indiscriminate and that the cause of the terrorist and victim are one. Even government efforts to target only those who participate directly in terror attacks—as when Israeli units, with advance intelligence of an imminent attack, strike terror operatives in so-called ticking bomb operations—might not be viewed by the community as a focused governmental reaction to threat when the targeted individuals are sons and daughters of the community.[14] It is not surprising then that public opinion polls in 2004 indicate that, despite worsening conditions brought about by Israeli retaliatory attacks, the majority of Palestinians supported suicide bombings against Israeli targets and that these levels of support rose precipitously at points during the second intifada (Bloom 2005: 193; Cordesman 2005: 382; Shikaki 2006). It is also predictable that civilian deaths from US aerial attacks in early 2006 on alleged terror targets in the tribal regions of western Pakistan provoked a strong, negative response from the local population and placed the Pakistani government on the defensive for its complicity in the US campaign (White and Khan 2006). A plausible consequence, then, of a government's war on terrorism is a hardening of attitudes, growing support for terrorists and their cause, and an increase in anti-government violence.

The challenges for offensive strategies arguably worsen when terror groups vie with others for domestic or international support. A group might try to increase its relative standing within a community—hoping to be seen as both capable and principled compared to other groups—through acts of extreme violence. The result is an increase in anti-government conflict and a raising of the bar for other groups that engage the government, which leads further to violence. Even Fatah (the "mainstream" Palestinian party within the PLO) formed its own secretive terror organization (Black September), which was responsible for the killing of Israeli Olympic athletes in Munich in 1972, after the radical Popular Front for the Liberation of Palestine hijacked and destroyed four airliners (three, in Jordan) in the notorious incident that sparked the Jordanian civil war. The same pattern was apparent years later when secular groups adopted the suicide bombing tactics of Islamic groups in the Israeli-occupied territories. In 1994, Hamas and Islamic Jihad started using suicide attacks against Israel; by 2000, secular groups such as the al-Aqsa Martyrs' Brigades and the Popular Front for the Liberation of Palestine were also employing these methods (Pape 2005: 20). Competition was apparent even in those attacks: Hamas initially foreswore the use of female suicide bombers but relented when they proved effective killing instruments (Harrison 2006: 197). A consequence of competition is that ceasefires within the Israeli-occupied territories tended to include all groups or no groups, and were generally short-lived. It was too tempting for groups to defect from a ceasefire agreement to reap rewards from challenging the status quo and from skewering rivals for their lack of sincerity and commitment to the cause. Indeed, Palestinian groups that chose the path of violence traditionally received a larger "market share" in the form of increased popularity (Bloom 2005: 24–5).[15]

With or without inter-group competition, the line of argument that retaliatory actions are self-defeating draws inspiration from the so-called action-reaction (or "spiral") model of international politics. In its simple form, the model assumes that actions beget counter-actions and that counter-actions beget counter-counter-actions, and so on and so forth, resulting in a spiraling of conflict out of control. This model inspires the common lament that participants in the Israel–Palestine conflict are locked in a "cycle of violence." This is a compelling notion if only because it converges with a widespread sentiment (grounded frequently in liberal assumptions) that conflicts are based on misperception and misunderstanding; it appears fairminded in not blaming just one participant; and it suggests, in the end, that all participants are worse off for their participation in conflict than they would otherwise have been. But it is also a problematic model, for a number of reasons.

First, the evidence is never conclusive that a cycle of violence is in place. The fact is that any sequence of battles can be interpreted to fit an action-reaction model—even if the conflict is entirely one-sided, with one party consistently challenging the other (acting) and, the second, consistently defending (reacting). In actuality, an action-reaction process can be quite irregular and asymmetric in appearance. Participants do not necessarily respond to conflict in kind, at set intervals: they act in keeping with their capabilities, in anticipation that others will act, with delay after preparing a response, and disproportionately to an attack to convince an opponent to capitulate.

Second, the notion of cycles is frequently employed disingenuously by those who want to *blame their opponent* for feeding the cycle. The cycle does not assume that one participant in the conflict is more moral, rational, or responsible than the other; the cycle—by definition—privileges neither participant. If it did, the cycle of violence would really be only a semi-cycle and could not be self-perpetuating. Indeed, to lament being victimized by a cycle is to suggest a perspective on the process that also violates the model's assumptions: a party that understands that the violence serves no useful purpose should be able to stop the violence (assuming that it is not simply an observer or innocent victim). That Hamas hid behind the rhetoric of the cycle is apparent when Hamas announced ceasefires, while rearming, so as to place the onus on Israel for violating the peace (Cordesman 2005: 204) and from Hamas' efforts to play a spoiler role in Israel–Palestinian negotiations. Kydd and Walter (2002) establish that Hamas' actions in the 1990s follow a global pattern in which extremists employ violence to derail peace talks by undermining moderates on both sides of the negotiating table. By eroding the opposing government's trust in its moderate negotiating partner (the Palestinian Authority) and perhaps even bringing a hard-line government leader to power (e.g. Israel's Ariel Sharon), extremists hope to induce immoderate government attitudes and policies that validate the extremist message, expose the traitorousness or naivety of the moderate group, and kill the chances of a peace agreement that will freeze out the extremists. From 1993 to 2001, then, Hamas focused its actions around six important events: the 1993 Oslo I agreement, the 1994 signings of the Cairo agreement (which established the Palestinian Authority) and Israel–Jordan peace treaty, the 1996 Israel and Palestinian elections, the 1998 Wye agreement (which called for the handover of land to the Palestinian Authority, movement toward a final agreement, and controlling of terror violence), and the 2001 Israeli elections that unseated the liberal Ehud Barak government in favor of a right-wing government headed by Sharon (Kydd and Walter 2002: 280). Counter to the predictions of the action-reaction model, Hamas conflicted in response to cooperation, not conflict. When Hamas did respond to conflict, it was conflict from "domestic" competitors. This was the case when Hamas used the pretext of an explosion that killed twenty people at a Hamas military parade in September 2005 as justification for a series of Hamas "retaliatory" rocket attacks on Israel. Hamas appears to have been pressured to act to deflect criticism by Palestinian officials of the military parades that Hamas had pledged would end; and its political predicament worsened when the Palestinian Authority called Hamas on its deception—asserting that Hamas needs "to shoulder its responsibility" rather than "making accusations against others" (el Deeb 2005). Whether Hamas was seeking to redirect anger for its actions or to capitalize on a political opportunity that the accident had created is less important than that Hamas felt its political predicament required a violent, external response.[16]

The cycle of violence argument is impugned further by a statistical analysis of Israeli–Palestinian violence in the September 2000–January 2005 period that took the lives of 3,200 Palestinians and 1,000 Israelis.[17] Jaeger and Paserman (2005) establish that the actions between Israel and the Palestinians fail to conform to a classic pattern of "tit-for-tat" reciprocity in which each side responds to the other with attacks of equal or greater ferocity (as measured by fatalities). Even if Palestinian groups were motivated to respond to violence with violence, they were arguably: (a) impaired by Israeli (offensive and defensive) actions from launching effective strikes; and/or (b) motivated to attack as well during (perhaps, because of) lulls in Israeli activity. Either possibility is inconsistent with the simple supposition that violence breeds violence—that motives and behavior hinge on the actions of an adversary and that levels of conflict spiral upward. Ironically, Jaeger and Passerman establish that the cycle argument appears to fit *Israeli* better than Palestinian behavior: the statistical pattern is for Israel to respond by inflicting fatalities when having weathered a fatal attack.[18]

Third, the focus of such cycles is typically based on a *study* of short-term effects with *assumptions* made about negative long-term effects.

Instead of spreading and exploding, conflicts can run out of steam when the public acts to constrain indiscriminate violence. In fact, the public has turned against terrorists who were once regarded as heros. The leftist Tupamaros of Uruguay enjoyed significant initial success in their efforts to rally the public and provoke government repression through a massive terror campaign (that included the kidnaping of high-profile persons). Nevertheless, the public turned against the Tupamaros and elected an unabashedly repressive, right-wing government to power that eventually eradicated the terror group (White 2002: 66–80). Even when governmental violence against a local population is perpetrated in the name of combating terrorism, it is by no means inevitable that the result of attacking terrorist staging bases and support networks is an inflamed public and escalating violence. Sometimes, populations (including Palestinians during the second intifada) turned their ire on terror groups for bringing retributive (Israeli) violence to the community (Cordesman 2005: 203), or for betraying their goals. The Tupamaros, PKK, Shining Path, Tamil Tigers, Irish Republican Army, and various Palestinian groups lost ground in the battle for the "hearts and minds" of a populous when targeting their constituency for extortion and thievery or engaging in political intimidation, infighting, or violent turf battles with ideologically like-minded groups. Testifying to this is the revulsion evoked within the Islamic community toward Abu Musab al-Zarqawi's organization in response to its suicide bombing attacks on Shiite civilians and beheading of captives in Iraq. Importantly, too, the cycle has been broken with the subjugating of populations through governmental violence. The brutal campaigns against dissent by the Baath governments of Syria and Iraq were followed by years of relative political stability within these countries. Indeed, the cycle has been broken despite dramatic escalations in hostility like that provoked toward Israel by its assassination of Yassin in 2004. Although tens of thousands of Palestinians witnessed or joined his funeral procession and Hamas vowed to avenge the killing—"to shake Israel like an earthquake" (CNN 2004)—Hamas joined the moratorium on attacks on Israeli targets in 2005 that preceded Israel's withdrawal from Gaza.

More generally, there is evidence that public enthusiasm for terror action can wane considerably when this support comes with a price. Polls conducted after September 11 and the US invasion of Iraq in 2003 established that bin Laden enjoyed substantial support throughout the Islamic world, where majorities of people in a variety of countries voiced the opinion that suicide terror attacks on US and Western civilian targets can be justified in defense of Islam. When the political, economic, and human costs of terrorism hit home—with spectacular terrorist attacks in Turkey, Indonesia, and Morocco—public support for Islamic terrorism declined. In a Pew research poll conducted on 6,200 people in 2005 in the primarily Muslim countries of Indonesia, Morocco, Pakistan, Turkey, Lebanon, and Jordan, evidence of an attitudinal shift was apparent: national populations continued to harbor resentment against the US, but far fewer people in most of these countries professed trust in bin Laden and support for suicide bombings.[19] Noteworthy in this regard is the dramatic change in sentiment toward al-Zarqawi's organization in Jordan once it was linked to a deadly suicide bombing attack at a hotel wedding ceremony in Amman in November 2005 (Gerges 2005). A large majority of the Jordanian public identified it, in consequence, as a terrorist organization. Also noteworthy is that, despite an aggressive Israeli counter-terror campaign that caused major economic and social disruptions and produced many civilian deaths within the occupied territories, support within Gaza and the West Bank for the suicide bombing campaign dropped with time. Whereas polls in the third quarter of 2004 indicated that roughly two-thirds to three-quarters of the Palestinian population supported suicide attacks on Israeli targets, such support was found roughly in a third to half of those surveyed by the second quarter of 2005.[20] These same polls revealed public optimism about the chances of a negotiated compromise between Israel and the Palestinians.[21] Whether or not the softening of attitudes is short-lived, it is hard to reconcile it with the action-reaction model, which predicts, instead, that the public will reject the voices of accommodation, view the gestures of an adversary in the harshest light, and see escalation as the only means to breach the impasse. It should be noted further that the attitudinal turnabout occurred without discernable evidence that the Sharon government would have been more forthcoming than was Barak (his left-of-center predecessor) at the unsuccessful 2000 Camp David summit negotiations that sparked the second intifada. Indeed, the right of Palestinian refugees to resettle in Israel—the issue that appears to have motivated Arafat's rejection of the Clinton proposals—is not one on which the Israeli government is likely to be forthcoming.

This is not to deny that social, psychological, and political processes create a self-perpetuating problem for governments that choose to "fight fire with fire." Under some conditions, the public is more inclined to blame the government for its counter-terror actions than terror groups for provoking those actions. These groups can profit from their image as modern-day "Robin Hoods"—a heroic image that derives from promoting lofty ideals and standing up to overwhelming government force. The image could survive despite the group's unheroic behavior. Given ethnic or religious ties to a local population, the group can engage in acts that create a public outcry when performed by outsiders. Iraqi insurgents have fired at US troops from mosques and used them to store weapons and ammunition, yet US troops are the ones that are accused of desecration when returning fire or moving to clear mosques of hostile personnel (Hassner 2006). Should these groups lose their romantic appeal, they can still count on a sociological influence—the legendary "Stockholm syndrome," so named for the city in which bank robbers held hostages for a period of days in a stand-off with police. Deprived of basic human comforts, subject to extreme stress, beholden directly to the robbers for life and any comforts extended, and pressed by outsiders (the police), the hostages saw the robbers as the "good guys" and the police as the "bad guys." They could not put their faith in outsiders who could only be "seen" through their provocations and always seemed to make matters worse. Indeed, these groups can use violence to political advantage. Writing on Iraq, Hoffman (2006: 113–14) notes that, "through violence and bloodshed, the insurgent seeks to foment a climate of fear by demonstrating the authorities' inability to maintain order and thus highlight their weakness." The strategy can work because "a population will give its allegiance to the side that will best protect it." Ironically, then, polls indicate that the US is blamed by the Iraqi public for the post-occupational hardships (or, at least, for not lifting hardships), including the insurgency. Blaming the US is a predictable response when the reality of life clashes severely with the promise that came with the US invasion.

In this regard, the US lessons in Iraq recapitulate lessons learned by other countries, including Israel in Lebanon: a "war on terrorism" fought by an outsider places it at a competitive disadvantage. Just as the US enjoyed initial success in unseating the Saddam Hussein regime, Israel succeeded early when it invaded Lebanon in 1982: the PLO was dislodged from the country and would no longer use it as a base of operations against Israel. But the Shiite population turned on Israeli forces after having first welcomed them as a counterweight to the Palestinians. With Shiite backing, Hezbollah eventually replaced Palestinian guerillas as the premier security threat along Israel's northern border and would offer instruction and support to Palestinian groups operating in the occupied territories during the second intifada.

Thus, public responses confound efforts to implement offensive-punishment and denial strategies. The ability of terror groups to draw from its support base and to energize that base, when under attack, severely challenge governmental efforts to defeat or punish those groups. The dilemma for the government is that it might not possess the will or capability— perhaps, brutal instincts—to overwhelm a terror opponent; at the same time, less-than-decisive force—in limited retaliatory attacks—can provoke the public and abet the terrorist's cause (on this trade-off, see Rosendorff and Sandler 2004).

Perspective three: organizational elements

The problems of impeding terrorist action through punishment and denial strategies persist when terrorists are recognized to operate as parts of complex organizations. Challenges exist whether these organizations are understood to be vertically structured—so that anti-terror strategies center on organizational "decapitation"—or horizontally structured—so that anti-terror strategies focus on disrupting the terror "network."

Vertical approaches: decapitating terror groups

A critical goal of US (denial-based) offensive policies is to disrupt and disable terror operations by targeting the al-Qaeda leadership. Such policies are good politics. Putting an ugly face on an abstract nemesis—presenting Germany as Hitler, Japan as Tojo, Russia as Stalin, Iraq as Saddam Hussein, and Islamic terrorism as bin Laden—can help mobilize public support for a sustained and costly war effort. But these efforts are problematic when applied to a war on terror. The nature and identity of the "leadership" of a terror organization is frequently subject to dispute and can change under attack to enhance the survivability and capabilities of the organization.

The historical record speaks to the effectiveness of decapitating a hierarchical terror organization. Despite a brutal campaign against the Peruvian state that cost the lives of tens of thousands of people in the 1980s, the Maoist Sendero Luminoso (Shining Path) movement in Peru was dealt a near mortal blow with the arrest of its leader, Abimael Guzman, along with a number of his compatriots in the country's capital in 1992. Thousands of guerillas surrendered to the government when promised amnesty in 1994, and the few who remained in hiding were confined to the remote jungles of eastern Peru. Attempts to revive the organization proved unsuccessful and were hurt further by the arrest of its new leader (Oscar Ramirez) in a military operation in 1999. Efforts to disable organizations from the top have been successful elsewhere. In 1999, Turkey's capture of Abdullah Ocalan was a major defeat for the PKK. The highly centralized party was effectively a "cult of the personality" around Ocalan. Consequently, Ocalan's arrest fractured the party through infighting, a loss in morale, and declining membership—in part because of Ocalan's own behavior in captivity. He disassociated himself from the actions of his followers and called upon them to stop fighting and to withdraw from Turkey. Subsequently, the PKK declared a ceasefire that lasted several years and withdrew its forces into Northern Iraq (Radu 2001).

Evidence in support of leadership targeting is also found on the Israel–Palestine front. The fundamentalist Islamic Jihad organization has remained active in operations against Israel in the occupied territories, but has not returned to the prominence it enjoyed when its founder and leader Fathi Shaqaqi was assassinated in Malta in 1995 (supposedly by Israeli agents) and replaced then by a leader who lacked Shaqaqi's personal magnetism and organizational capabilities. Likewise, Hamas was hurt by the targeting of its leadership in a relentless Israel campaign that, among others, took the life of Sheikh Yassin. (Hamas was targeted in half of the 159 Israeli assassination attempts in the September 2000–April 2004 period, compared to seventeen percent that targeted Islamic Jihad; see Zussman and Zussman 2005.) Hamas suffered at least to the point that its political leaders pressed Israel to stop the assassinations of militant leaders (David 2003: 143–4) and, with the killings of Yassin and his successor, Abdel Aziz Rantisi, adopted the furtive lifestyle of Hamas' military leadership and refused even to reveal Rantisi's replacement (Myre 2004). By itself, increased secrecy imposes costs on an organization. It requires that leaders reduce their profile, which makes it harder for them to rally the public (as Yassin and Rantisi had done) to support the cause. Moreover, secrecy brings structural modifications, distractions, and resource expenditures to the organization that compromise its efficiency (see Chapter 6): "to avoid elimination, the terrorists must constantly change locations, keep those locations secret, and keep their heads down, all of which reduces the flow of information in their organization and makes internal communications problematic and dangerous" (Byman 2006: 104).

In the main, though, decapitation is neither easy nor effective. For various reasons, attacks on obvious targets—offices or private homes—might not bring down a government, let alone a terrorist organization. First, leaders can be evasive. Israel had some success in targeting the terror leadership: twenty-five percent of Israel's assassination attempts were aimed at organizational leaders, with an eighty-five percent rate of success (Zussman and Zussman 2005). But Israel's position is unique in that the leaders of Hamas—a principal target of assassinations—maintained a relatively visible presence within a geographically confined area. Rantisi was killed, for example, in his car after leaving his family home in Gaza (Myre 2004). Then, Israel benefitted in targeting from its elaborate intelligence network and its ability to bring military forces to bear quickly when targeting opportunities arose (Byman 2006: 100). Without these advantages, terrorist leaders have been able to hide and operate on the run. Bin Laden is certainly a case in point. He has escaped capture and death in Afghanistan (and its border regions) despite his global reputation, physical stature and recognizable features, a huge price on his head, and substantial US intelligence and military assets devoted to the hunt. The unsuccessful effort resembles that to capture the warlord, Mohamed Aideed and his aides, during the UN-backed operation in Somalia. An ensuing military confrontation in Mogadishu (the infamous "Blackhawk down" incident of 1993) set in motion by a failed US attack on a leadership conclave sparked an unceremonious US withdrawal from the country. Second, leaders have entrenched themselves firmly in positions of power by eliminating and intimidating their opponents. Thus, anything less than a "fatal shot" might not remove a leader from power. The US "shock and awe" bombing campaign of Baghdad in 2003, aimed at the power centers of the Iraqi Baath regime, did not bring down the Iraqi government. Contrary to optimistic scenarios, opponents did not rise up against Hussein once the US made its military presence felt. Instead, the Iraqi regime had to be physically driven from power by US ground forces. Third, organizations manage to survive the loss of their leaders. The long-term effectiveness of efforts to decapitate an organization reflects its reliance upon the exceptional martial capabilities, organizational skills, or personal charisma of a leader. But many organizations are not that reliant upon a single leader. Hamas has proven resilient even with its large leadership "turnover." Moreover, the consensus among US and Iraqi officials is that the Iraqi insurgency is ad hoc and horizontal in structure, notwithstanding the public US focus on the battle against al-Zarqawi in Iraq (Filkins 2005).

Evidence supporting decapitation tactics looks less impressive under scrutiny. The arrest of Ocalan coincided with waning support for the PKK within Turkey's Kurdish population, the critical loss of Syria as a PKK base of operations, and a Turkish military push into Northern Iraq. For its part, Islamic Jihad eventually did recover from the Shaqaqi assassination under new leadership to contribute significantly to the suicide bombing campaigns of the second intifada. Nor can it be said that leadership targeting turned the tide for Israel in its battle against terrorist attacks. The numbers appear to speak to the military success of the Israeli effort: 55 people died in 12 bombings in 2004, compared to 228 people who died in 42 suicide bombings in 2002 (Eggen and Wilson 2005). But Israel's targeted assassinations were paired with aggressive efforts to disable bomb-making factories and destroy weapons caches, pervasive roadblocks, curfew enforcement, travel permits to encumber public movement, an alert and mobilized Israeli citizenry in areas that were vulnerable to attack, an elaborate intelligence network to forewarn of imminent attacks, and a security "fence" that separated Israel (and territories claimed by Israel) from the rest of Gaza and the West Bank.[22] Furthermore, these favorable numbers occurred despite an increase in the number of Palestinian attacks (see Byman 2006: 103).

Apart from their effectiveness, decapitation tactics risk wide-ranging, negative political fallout. These effects occur when targeted assassinations kill the "wrong" people (with or without the "right" people);[23] embarrass or compromise foreign governments (when the killing occurs on their territory or with their assets, e.g. passports); create martyrs for the cause or make heros of those who survive repeated assassination attempts;[24] kill "moderate" leaders who are open to negotiations and exert a constraining influence on organizational violence; reinforce a public belief that the attacking party is brutal and unscrupulous and bring demands for revenge; or inspire organizational changes that reduce the decapitation threat. Indeed, a single killing could have a number of untoward effects. For instance, the 1996 killing by Israeli agents of Yahiya Ayyash, Hamas' chief bomb-maker (aka the "Engineer"), who was reputedly linked to hundreds of civilian deaths, provoked a wave of terror against Israeli targets and a decision by Hamas to train and disperse bomb-makers throughout the West Bank and Gaza (Graeme *et al.* 2004: 185–6). Thus, a consequence of a successful killing could be a robust and virulent terror organization. Islamic Jihad became far-less dependent on a single leader and resembles other groups that are "so loose in their organization that true decapitation is no longer possible" (Byman 2006: 100). Likewise, mass arrests of the Hamas leadership by Israel in the late 1980s and early 1990s inspired Hamas to adopt an organizational structure that relies upon geographically separated, secretive, self-contained cells and to divide and disperse its leadership to Amman, London, and the suburbs of Washington, DC (Mishal and Sela 2000: 56–8). These reforms hurt the efficiency of the organization but also efforts to destroy it, and to moderate its behavior. The expatriate leaders of Hamas have been less inclined than "internal" leaders to concede to political and military "realities on the ground" that push toward compromise (Mishal and Sela 2000: 151–69).

Attacking terror organizations from the top can produce a pernicious result when the remnants of the organization are radicalized under new leaders that seek to avenge the past. There is evidence, for instance, that retaliation for the killing of Hamas's chief bomb-maker was conducted by maverick cells linked to Ayyash, on a campaign of personal revenge (see Gunning 2004: 242). This decapitation problem was familiar to Cold War strategic thinkers who recognized that attacking central command and communications centers can backfire if the parts then act independently of the whole. A further danger exists, however, when surviving individuals and groups compete for prominence within the movement through bigger and bolder terror operations. Then, decapitation tactics would be undermined both by the independence and the interaction of the parts.[25]

These arguments do not strengthen the case for designing an offensive strategy around beheading the al-Qaeda leadership; nor does the statistical evidence. According to the Bush administration, almost three-quarters of al-Qaeda's September 11 leadership is now dead or imprisoned; yet the global "terror threat" does not appear to have decreased in magnitude. In 2004, for instance, 3,192 terror attacks were said, in one US government report, to have occurred around the world resulting in 28,433 wounded or dead (*Washington Post* 2005). In 2005, 11,111 attacks were reported by the same source to have killed over 14,000 people (DeYoung 2006). The numbers from Iraq are no more heartening. Despite the US optimism that the killing of Abu Musab al-Zarqawi in June 2006 would reduce the violence in that country, the Iraqi death toll from terror attacks soon surpassed an average of 100 people a day (see e.g. Semple 2006). These high numbers speak to the limits of trying to decapitate a menace that can easily grow a new head, has many heads, or can arise spontaneously in a new location.

The relentless pursuit of al-Qaeda leaders appears actually to have trans-formed their organization. Bin Laden and other al-Qaeda leaders still rally the faithful for a holy war against Christians, Jews, and their cronies, and al-Qaeda operatives continue to disseminate their advice and attack tech-niques around the world through the Internet and other sources. But al-Qaeda is now less a coherent, centralized adversary and more a symbol, model, and inspirational trademark that is used by groups capable of inde-pendent action. These groups have attacked political and economic targets around the world by relying upon low-cost, unproblematic, and effective killing technologies and a relatively uncomplicated command structure. The evidence does not suggest, for instance, that the Moroccan radical group that carried out the Madrid attacks in 2004 that killed almost 200 people, or the London subway attacks that killed over fifty people in 2005 were organizationally linked to "al-Qaeda," as traditionally understood. In fact, the bomb attacks in Madrid (like the one in Morocco) combined the efforts of external leaders trained abroad and locals with clean records who were not likely to draw unwanted attention from police and intelligence units prior to the attack (Frankel 2005).[26] The extent to which some recent terrorist operations involved external links remains a matter of dispute. These operations include the London attacks and the October 2004 Taba and July 2005 Sharm el-Sheikh bombings that targeted the tourist industry in the Sinai (killing more than 100 Egyptians and foreigners). What is not in question is that the attacks were carried out by local elements with local resources.[27]

In sum, there is considerable evidence that targeted killings can disable a terror organization. Decapitation tactics are bound to impede the produc-tive uses and flows of goods, people, ideas, and resources within a hierarchic, terror organization, which makes it less efficient; and the coor-dination problems among operating parts of a decapitated organization can prevent it from capitalizing on advantages of scale or reach. There is also reason to believe that leadership targeting can hurt terror organizations, such as al-Qaeda, despite a more diffuse organizational structure. The killing of bin Laden would deprive global terrorism of its most visible and tireless proponent and amount, at the very least, to a symbolic blow to radical Islamic groups that have united under the al-Qaeda banner. Decapitation could have important deterrent effects, too, should terrorist leaders seek to protect their organizations (even if not themselves per se) by acting secretively and evasively (thereby compromising their organiza-tion's efficiency), or by holding their fire so as not to provoke retaliation. This was certainly the hope when, in attempting to get Hamas to cease its rocket attacks on Israel in late 2005, Israel's Defense Minister issued a not-so-thinly veiled decapitation threat—naming two current Hamas leaders and threatening to send them "to where Abdel Aziz Rantisi and Sheik Yassin are" (quoted in Wilson 2005b).

But the liabilities of decapitation tactics also remain considerable. Efforts to kill terrorist leaders might not succeed; and even a successful hit might not prevent a terror groups from accomplishing its goals. Indeed, decapitation tactics can produce a horizontal restructuring of the terror group that makes its leaders more difficult to track and kill and, more generally, that disperses organizational assets to make them difficult to locate and destroy. Then, these tactics will lack coercive punch. The threat of decapitation is less effective against leaders that are hard to find or are lost in the crowd of individuals who assume control of a decapitated terror group. The more leaders, the less the value to the decapitating party of the life of any one, the lower the vulnerability of any one (if others can draw fire), and the stronger the motivation for each leader to act out with violence to compete for prominence within the organization.

Horizontal approaches: attacking terror networks

A vertical approach to combating and punishing terrorists requires an understanding of terrorist organizations as hierarchic in structure and adopts the efficient approach of disabling terrorist organizations from the top. In comparison, a horizontal approach requires an understanding of terrorist organizations as "networks" and adopts the efficient approach of attacking their weakest links. This can be accomplished by destroying those links (a denial strategy) or holding them accountable for their behavior (a punishment strategy). Indeed, the horizontal approach is arguably an extension of the vertical approach when skilled operatives and key functionaries, not the top leaders of an organization, are targeted.

Increasingly, global terrorism is understood as a "network" (Ronfeldt and Arquilla 2001) characterized by interconnected nodes (e.g. businesses, terror cells, terrorist leaders) and their redundant linkages (physical and otherwise). Al-Qaeda itself has been described as a "network of networks" in which "responsibility and decision-making authority are devolved down to the lowest possible level," "each node is unaware of the identities and attributes of others," and linkages are maintained with still other networks (Deibert and Stein 2002: 7). In fact, some have noted that al-Qaeda and other terror groups, as "violent, secretive organizations," share many features with criminal organizations:

> their cellular and networked structures extending across national boundaries; their high level of energy, fed by sentiments of revenge; their sponsorship by states or elements of states; their parasitic revenue streams from licit and illicit commerce; and their tendency toward extraordinary violence in some historical moments.
>
> (Schneider and Schneider 2002)

What is clear too is that the Internet has embellished the horizontal features of the global terrorist network by allowing terrorists to operate without central direction. The web has helped terrorists to recruit: via the web, terror leaders can proselytize to a global audience through religious and ideological appeals and the glorification of jihadist exploits with videotape of attacks and tales of martyrdom. The web has empowered terrorists, as well, by providing a ready source of instructive manuals, explosive recipes, and logistic support, for example, in the form of information on possible travel routes, cover stories, and sanctuaries for operatives. In addition, the web has proven a potent device for communication among affiliated individuals and groups. For instance, Iraqi insurgents have used bulletin boards to offer tactical advice to operatives in other parts of the world and have used the Internet to generate external support for operations in Iraq. The negotiations that led to the merger between the forces of Abu Masab al-Zarqawi in Iraq with al-Qaeda were conducted through the Internet. Benefits of the Internet are that it has limited the need for personnel to gather, fostered linkages between individuals at every level of command, and effectively "cut out the middleman" so that individuals can now prepare for terror operations without the aid of instructors and training camps. In consequence, the Internet has extended the global reach of the terror enterprise by flattening and dispersing the global terror structure (Coll and Glasser 2005b, 2005c).

By recognizing the existence of such complex networks, and determining and exploiting their essential points of vulnerability, a horizontal attack strategy aims to impede the effective functioning of complex organizations through disabling attacks or coercion. As Davis and Jenkins (2002: xi) argue in support of a coercive strategy:

> it is a mistake to think of influencing al-Qaeda as though it were a single entity: rather, the targets of U.S. influence are the many elements of the al-Qaeda *system*, which comprises leaders, lieutenants, financiers, logisticians and other facilitators, foot soldiers, recruiters, supporting population segments, and religious or otherwise ideological figures. A particular leader may not be easily deterrable, but other elements of the system (e.g. state supporters or wealthy financiers living the good life while supporting al-Qaeda in the shadows) may be.

The key, then, to denial and punishment strategies against terrorist organizations is identifying those parts of the network that will most hurt it when removed (i.e. impairing critical functions or multiplying negative effects), at acceptable cost to the attacking party. Thus, sensitivity to the requirements of a horizontal approach is apparent when states target groups aligned with more dangerous groups, confiscate assets used to finance terrorist activities, develop safeguards to prevent dangerous chemicals and materials (e.g. used to manufacture explosives) from falling into the wrong hands, dismantle buildings used as safe houses or bases for terror operations, assassinate mid-level terrorist leaders who possess field knowledge and operational expertise, and so on and so forth.

Evidence validating the horizontal approach is arguably offered by the US success in depriving al-Qaeda of its primary base in Afghanistan for operational planning, training recruits, and financing operations and the less-conspicuous accomplishments of US special operations forces in training and supporting Latin American, African, and Asian foreign military units to police their own territories. Further validation is found in efforts to combat indigenous terror movements by targeting their support structure. For instance, Israel achieved considerable success in reducing terror attacks on Israeli territory through a massive, forward military presence in Gaza and the West Bank. This allowed Israel to control key trafficking points (e.g. between Gaza and the Sinai), create observations points and separation zones, block access routes, cultivate intelligence sources, and engage Palestinian units that could conduct terror attacks. The level of terror violence subsided dramatically and immediately, in fact, with Israel's 2002 offensive thrust into the West Bank: the number of successful terror attacks and overall casualties in attacks on Israeli targets (in Israel and the occupied territories) never again came close to their prior monthly levels.[28]

The reality though is that it is difficult to disable or disrupt networks, whether these involve drug smuggling, trafficking in contraband, illegal immigration, or terrorist activities (Lepgold 1998), or whether the attack strategy is based on principles of denial or punishment. The reasons are as follows.

First, these networks are "opaque" in the sense that, to outsiders, the function and importance of parts of the network are unclear. It is hard to follow the money trail through networks of charities (that might also fund education and medical services), legitimate and illegitimate business enterprises, front groups, and banks that have much to lose financially through rigorous scrutiny of clientele and are left to police themselves. Some parts of networks are entirely shielded from scrutiny. This is true of terrorist operations within so-called international black spots, where terror groups can surreptitiously coordinate their activities and command their networks from territories under their control (e.g. the portions of Colombia controlled by the leftist Revolutionary Armed Forces of Colombia, or FARC) or effectively under no one's control (e.g. al-Qaeda operations in the lawless tribal regions of western Pakistan). The network's electronic components are also well masked. Organizers have hidden their activities by establishing on the web—to quote one expert—"a more anonymous, more protected, more nomadic presence"—as fixed addresses have rendered terrorist websites vulnerable to monitoring and disruption (Coll and Glasser 2005b). For instance, terrorists communicate by posting messages on discussion boards, hiding material directed at a single user in spam e-mail sent indiscriminately to a large number of users, and saving e-mail messages in draft form in accounts (from free and anonymous e-mail services) and then communicating the account name and password via other forms of Internet communication (Coll and Glasser 2005b). Financial networks are no more transparent. Terrorists that rely upon banks to transfer money often move their finances in small amounts that are less likely to draw attention. Banks cannot identify these as "terrorist transfers" without first knowing who the terrorists are. Terror groups have also relied upon cash in small, hand-carried bundles, ATM withdrawals of monies deposited in other countries, stored-value cards (e.g. "gift cards"), and informal "hawala" networks (Slevin 2006). This trust-based, remittance system is a traditional means for transferring money off the books to, from, and within the Middle East. It allows cash provided by a "sender" in one part of the world to be "received" in another without an actual "transfer" or official records. The system operates through faith that the money given to the hawala agent with which the sender interacts is properly connected to the hawala agent with which the recipient interacts, so that money given at one place will be paid out at another. It does not require that either the sender nor recipient disclose their identities, only that the parties know a password.

Second, these networks are "elusive" in the sense that targeted parts of the network can move, perhaps to return after an attack. This is certainly apparent in the military sense. Attempts to root out Iraqi insurgents in certain hotbeds of resistance were stymied when insurgents shifted their operations to other cities or resurfaced in prior locations when the US offensive ended (Graham 2005).[29] The electronic parts of the network are more elusive. For instance, terror websites used for recruiting and communication have been closed down by authorities only to reappear under new addresses; and material posted by terror groups to bulletin boards is quickly linked to other jihadist sites (Coll and Glasser 2005c). The financial parts of the network are equally slippery, as revealed by problematic efforts to compile central lists of those who finance terrorism. Through an act of the UN Security Council, member states have been directed to adopt appropriate regulatory action, criminalize terrorist financing, and freeze assets of entities that the Council lists as al-Qaeda affiliated. The list contains names of business enterprises that no longer exist, operate under a variety of names, and act in silent partnership with organizations involved in "legitimate" enterprises.

Third, these "networks" are not necessarily networks: terror organizations are often self-contained and capable of autonomous operation. Terror groups have become highly self-sufficient through involvement in criminal enterprises such as drug production and smuggling, trafficking in contraband, credit-card fraud, bank robberies, extortion, and kidnaping. It seems, for example, that the Iraqi insurgents have been funded substantially through the black market oil trade and the diversion of Iraqi government monies through corruption (Worth and Glanz 2006). Thus, efforts to destroy parts of the "network" might have, at best, a local effect but could have broader—and decidedly *negative*—effects if causing these parts to become networked. The Iraqi insurgency is armed and staffed overwhelmingly from within Iraq, with the goal of expelling US forces and promoting the interests of the Sunni minority. By attacking these insurgents, the US has increased their links to external groups—for that matter, the link between these and still other external groups (Hoffman 2006: 115). One effect, then, of US attempts to excise Iraq from the terror network has been to provide military training and ideological fervor to foreign operatives who return to their home countries (e.g. in Europe), to pass their skills and motivation to others. Another has been to inspire indigenous populations throughout the world to perform acts of terror by linking, via the Internet, with outsiders to obtain skills, motivation, and assistance to engage in terror attacks (Atran 2006: 141).

Fourth, these networks possess links that are "redundant" in that multiple connections between critical nodes permit alternative routes when other routes are blocked or disrupted. The advantages of networks lie in "their abilities to *shrink the distance* between any two nodes by finding efficient routes between them and to connect nodes through rerouting when a pathway is blocked" (Matthew and Shambaugh 2005: 619). Thus, with efforts to freeze the assets of terrorist organizations and monitor the transfer of money that could be used to finance terrorism, terrorist organizations have relied upon other fungible means for transporting wealth such as precious stones. They have also been able to protect their tangible assets and business enterprises. Whereas the US has spearheaded aggressive international action against money laundering and banks that do business with terrorists, shutting down businesses that front for terrorism and halting the smuggling, extortion, and property theft upon which terror groups thrive are severe challenges. Terrorist organizations have forged symbiotic relationships with criminal organizations (e.g. Colombia's FARC with Mexican drug traffickers)—capitalizing on their expertise in areas such as counterfeiting and money-laundering or obtaining logistic support for smuggling arms or personnel. In fact, they have transformed themselves into criminal organizations. That the Palestinian Abu Nidal organization (once affiliated with Fatah and responsible, in the 1980s, for a number of high-profile airline hijackings and attacks) virtually morphed into a criminal enterprise, that the Abu Sayyaf organization in the Philippines increasingly focused on hostage takings for ransom, and that FARC became deeply involved in all aspects of the Colombian drug trade speaks to a push by these organizations to become more self-sufficient, profitable, and resilient (at a cost, too, in their "noble" intent; see Makarenko 2004).

Finally, these networks are "diffuse" in that their capabilities are spread widely. In consequence, a coordinated and perhaps simultaneous international effort is required to prevent an organization from transferring its operations across borders into countries with lax or venal governments, poor enforcement capabilities, different legal or political principles, or a favorable disposition toward the group. Governments (e.g. Saudi Arabia) have not followed through on their pledges to take action against certain businesses and charities—refusing to act against organizations, such as Hamas, that are not "terrorist" organizations by those governments' definition. Western countries have also found it difficult to coordinate their efforts to track terror operatives and to bring terror suspects to justice. The sharing of intelligence information between the FBI and CIA has been problematic; it should come as no surprise that such sharing is encumbered across national borders. National intelligence agencies have held information close so as not to compromise intelligence assets even when the information was needed to prosecute terror suspects (in foreign and domestic courts) and prevent their release. The diffusion of the network has also hobbled efforts to combat terrorism on the financial front, allowing those identified as terrorist financiers to operate in the open.

One example is telling *because* it has received considerable attention. In 2005, a US television network (NBC) reported in its nightly newscast on NASCO, a Nigerian based company that was owned by Ahmed Idris Nasreddin who had been listed by the US and UN as a terrorist financier in 2002 and whose US banking assets were announced by Bush personally to have been seized back in 2001 (Myers et al. 2005). Consequently, some of his assets were frozen in Morocco, Turkey, and Italy but not in Nigeria where NASCO sponsors polo events and markets beauty products and cornflakes. These moves did not affect company stock, most of which is owned by a Panamanian company—apparently owned by Nasreddin. Moreover, Nasreddin still operates quite visibly in countries that have frozen some of his assets. In 2003, the same network reported that Nasreddin owned the Hotel Nasco which operates in Milan, Italy. Nasreddin continues to run his business network from Morocco and Campione d'Italia (Comras 2005).

In principle, it is possible to disrupt a network through punishment and denial strategies that *count on* interdependencies between parts of an operation. Then, many possible strategies can be employed singularly or simultaneously to disable an organization. But the problems of deterring and disabling terror networks are considerable, as expressed best in a UN report, "It will always be difficult to design, let alone enforce, sanctions against diverse groups of individuals who are not in one location, who can adopt different identities and who need no special equipment to launch their attack" (UN Security Council 2004). As a result, attacks directed against any one part of the network might leave more important parts of the network intact, only temporarily impair the operations of the organization, or inspire organizations to establish new links with external groups. The unfortunate reality is that much of the terror network is hidden and resilient and can survive attempts to disrupt its parts. In attacking it, then, an appropriate comparison is US attempts to encumber the North Vietnamese war effort through heavy and repeated bombing of the Ho Chi Minh trail—a network of jungle paths through which North Vietnamese forces and supplies moved by foot and bicycle, with increasing success, toward the south (Murray 2001). US weapons were too blunt, given the challenge, for the US effort to succeed.

In sum, the relatively low costs of participation in many important facets of the contemporary terror network (e.g. the internet) and the gains to be had from participation in others (e.g. illicit commerce) hinder a denial strategy intended to disable the network. If participants thereby conclude that they incur low costs and few risks in joining the terror network, coercion cannot multiply the disabling effects of force.

Conclusions

Whether the intent is denial, punishment, or some combination of the two, effective offensive strategies against terrorists are complicated by a variety of factors. These problems include targeting an elusive and impermanent terrorist infrastructure; the limited usefulness of retaliating against terrorists when popular support for terrorism can grow under attack and retaliatory actions validate the terrorist message; the questionable effectiveness of retaliating against state sponsors or hosts that can deny their complicity or are too weak politically and militarily to confront a terror group; and changes in the structure of an organization that could render it more potent. A vertical approach of dismantling terror organizations from the top is ineffective when leaders can be replaced and counterproductive when an organization can fragment into smaller groups (or compact terror cells). These groups might harbor a now stronger sense of grievance under a radical leadership that need no longer compromise to maintain organizational cohesion. The liabilities of a horizontal approach of disabling terror "networks" include the ability of the terror organization to reconstitute itself, adapt to the changing threat, learn to cover its tracks, operate autonomously, or develop new links to outside groups.

This does not mean that offensive strategies aimed at deterring or disabling terrorist groups will fail. Terror groups and opposing governments sometimes share interests that create a basis for compromise, which permits deterrence to work. Notable, for instance, is that most terror groups have neither the reach nor ambition to hit targets within the US and will avoid striking US targets when this could provoke a US response or waste precious organizational resources. Deterrence strategies can even perform against an implacably hostile terror foe. Governments have had some success in holding other governments accountable when their national soil is used by terror groups as a base (e.g. Jordan) or staging ground (e.g. Libya) for attacks, and in limiting the nature and degree of support that governments extend to terror groups. For example, Syria and Iran have abetted terror operations within proximate territories but have shied from sponsoring global terrorist campaigns. Strategies can work, as well, when intended primarily to fight rather than to deter terrorism, as evidenced by the success with which some governments have forcefully reduced levels of terror violence. Indeed, the principle that violence only begets violence does not appear to hold throughout much of the Middle East, a central focus of terrorist grievances. A paradox of the region is that its governments are among the world's most oppressive *and* entrenched in power. Contrary to assumptions at both ends of the ideological spectrum, there is anecdotal evidence suggesting that violently confronting terror groups can result in their defeat and evidence that it cannot. The statistical evidence is equally equivocal. Given the weak correlation worldwide between democracy and terror attacks, it seems that many autocratic countries have quashed violent opposition with repressive practices. Of note is that the Indian democracy appears to experience far more terror incidents than the more populous, nondemocracy of China (Gause 2005).

In the final analysis, it is important not to exaggerate the importance of any single experience by offering "lessons" without context or reference to the motives and strategies of a given terror group. Inevitably, too, success and failure must be judged by what governments sought to achieve through their counter-terror efforts and at what cost. For instance, any success that Israel obtained through aggressive, offensive policies came at a high price in financial and human resources for all parties to the conflict and ultimately owes, in part, to Israel's willingness to reduce its offensive burden by rendering concessions that could reinforce the opponent's belief that "terror pays." For that matter, Israel's success in lessening the terror threat was due in part to Palestinian success in convincing Ariel Sharon, Israel's Prime Minister, that occupation does not pay—that Israel could not afford the military, economic, political, and human costs of maintaining Jewish settlements in Gaza. Thus, Israel succeeded because it accepted a policy of containing rather than eliminating the terror threat.

In comparison, the US has fought its war on terrorism from a severe disadvantage: it has pursued its ambitious objective of "defeating" global terrorism by attacking widely from a distance, in geographical, social, political, and informational terms. In consequence, the war effort has been hampered by the US's limited capabilities to apply significant force in disparate locations; its insufficient understanding of local cultures, languages, and traditions; its poor reception worldwide given deep suspicions about US motives; its inability to distinguish terrorist from non-terrorist assets; its dependence upon "friendly" governments that are loath to cooperate on

US terms or too anxious to sell their domestic security efforts as part of the US war on terror; its reliance for information on those who distort the truth to serve their own political agendas or personal vendettas; and its reluctance to engage any adversary in costly and protracted conflicts. Such an effort is unsustainable, and any effort is doomed to fail without judicious US choices about where, when, and how to intervene. In particular, US policymakers must avoid policies that require the US to confront nationalist-oriented, terror groups on their own turf ("taking the war to the enemy"). These policies make it too easy for groups to evade the consequences of their actions through mobility and subterfuge, to play for time with the belief that the US will eventually exhaust itself, and to extend their support at home and abroad by portraying themselves as innocent victims of US aggression and corrupt and repressive US-supported regimes. This is not to deny the conditional effectiveness of offensive counter-terror strategies. But it is to acknowledge prohibitive liabilities when offensive strategies are meant to uproot a deeply imbedded terror infrastructure.

The US could benefit, then, by redirecting its offensive efforts toward striking groups when and where they emerge in preparing attacks (as discussed in Chapter 6). In keeping with the notion of a "proactive defense," the US can still vigorously acquire intelligence on the activities of terror groups, pursue and arrest terror suspects, and cooperate toward those ends with like-minded allies (in Europe). It can also operate from "forward" military positions and respond to threats before they translate into actual attacks; indeed, a primarily defensive stance does not proscribe full-scale US military actions against established and concentrated threats such as those posed by al-Qaeda and the Taliban in Afghanistan. But a proactive defensive strategy shifts the security focus toward "proximate" threats. It is not intended to "address," or to substitute for addressing, underlying grievances that cause people to become terrorists or to support their cause. In fact, it is best pursued with awareness that aggressive actions against groups with a nationalist or economic agenda can push these groups to identify, and seek common cause, with al-Qaeda or other global terror movements. Whether this strategy will succeed depends on whether the US is able to pick its battles, set reasonable goals, and understand that even obtaining these goals comes with a hefty price.

It should also be noted that governments have had some success in reducing the level of violence when offering positive incentives to terror groups to join the political process: a 2002 ceasefire held temporarily in Sri Lanka when its government offered to legalize the Tamil Tigers and negotiate with them as long as they kept the peace. More noteworthy is that the Irish Republican Army has moved toward disarmament and pursued its objectives (via its political wing, Sinn Fein) through an electoral strategy and that Hamas in the Palestinian territories, Hezbollah in Lebanon, and Sunni and Shiite groups in Iraq have limited their violence to some degree when opting for increased participation in governing institutions. Governments can also experience success if able to drive a wedge between terrorist groups and their public supporters by addressing their economic, political, and social grievances. In the wake of Israel's pullout of Gaza, polls indicated that residents of Gaza were more inclined to support the disarming of Palestinian militants than were residents of the West Bank, where Israel still maintained a military presence (Stephens 2006). Admittedly, the recommended mix of carrots and sticks remains unclear in the relatively large academic literature on civil conflict. The very factors that give people reason not to rebel can create motivation and opportunities to rebel (see Collier and Hoeffler 2001; Lebovic and Thompson 2006; Sorli *et al.* 2005: 145). Economic development can provide people with a better standard of living, but can also uproot them with a false promise of a better life and provide means for organizing and financing rebellions (on these issues, see Collier and Sambanis 2002; Davenport 1995: 692; Sambanis 2002: 220–1). Evidence from the Israeli-occupied territories suggests, at best, that there is no correlation between educational and economic levels that Palestinians obtain and their support for armed action against Israel (Krueger and Maleckova 2002). Impoverishment can actually dampen support for such violence. Germane, here, are the results of a recent opinion survey of fourteen Islamic countries that found that respondents who claimed that they could not afford to buy food were more likely to oppose terrorism than respondents who owned cell phones and computers (Fair and Haqqani 2006). Thus, government efforts to improve the lives of people will not necessarily reduce levels of civil conflict, and governments can arguably maintain domestic control through military measures—perhaps, repressive practices—when unable or unwilling to assuage grievances through distributional or compensatory policies.

Without rendering moral judgments, it cannot be said finally that there is only one path to combating a terror problem. But it can be said that the path of violence is fraught with peril and is not for the awkward or faint of heart.

6

A DEFENSIVE-DENIAL STRATEGY AGAINST TERRORISTS

Homeland security

The September 11, 2001 attacks on the World Trade Center in New York and Pentagon outside Washington, DC produced a profound sense of vulnerability within the US. To US government leaders, the evidence was now irrefutable that inflicting widespread death and destruction in the US had become key terrorist objectives, that terrorists would run high risks to implement these goals, and that the physical reach of these "new terrorists" was limited only by their ingenuity. They promised a future in which airliners loaded with fuel were transformed into firebombs, municipal water systems were poisoned, biological plagues were released on an unsuspecting population, and crude radioactive devices or nuclear bombs were exploded in populated areas.

Although alarming, the new breed of terrorist does not represent a wholesale break from the past. Like prior US nemeses, the new adversary does not simply lash out. The signatures of al-Qaeda attacks around the world are intense planning and coordination, active reconnaissance and intelligence gathering, long gestation periods (four, five, and six years in the attacks on the USS *Cole*, the US embassy bombings in Africa, and the September 11 attacks, respectively), and adaptation to changing threats, opportunities, and constraints.[1] If so, the attacker's willingness to withhold attack until conditions are right and forces are in place, to size the attack for operational success, and to change targets in the face of setbacks suggest that "terrorism is likely to be a reasonably informed choice among available alternatives" (Crenshaw 1990: 11). Consequently, terrorists can be deterred from carrying out attacks, at least in the short term and/or against certain targets. As Davis and Jenkins (2002: xii) observe, "the empirical record shows that even hardened terrorists dislike operational risks and may be deterred by uncertainty and risk." This is true down the chain of command: "terrorists recognize that their power depends on perceptions of whether they are winning or losing; their leaders are deeply concerned with *control*; and martyrdom in a stymied mission lacks the appeal of dying in a spectacular, successful attack." Put simply, operatives "may be willing to risk or give their lives, but not in futile attacks" (Davis and Jenkins 2002: 16).

That al-Qaeda has resorted to meticulous planning, then, is both chilling and heartening. The al-Qaeda surveillance reports for possible attacks on the World Bank, International Monetary Fund, and New York Stock Exchange that came to light in August 2004 include intricate detail on pedestrian and traffic patterns, timing and methods for gaining access to upper floors, the location of security cameras and personnel, and security routines. The reports reveal that the perpetrators had unencumbered access to the targeted facilities and an impressive ability to identify logistic problems, operational challenges, and potential solutions. Yet these same reports establish that operatives recognized their task was complex, that success hinged on the ability to perform a difficult sequence of tasks, and that the advisability of an attack depended, then, on its likely success. Here, as elsewhere, the attacker weighed its capabilities carefully to determine whether they were up to the tasks at hand; by implication, it understood that it must adjust, compromise, or wait if necessary.

From a *denial* standpoint, if an attack on a target at a given time will be ineffective, the attacker has no incentive to attack. Indeed, the attacker has strong reasons not to attack. When attacks fail, the attacker must accept costs (punishment) in the form of a lost capacity to attack alternative targets or the same target, at some future point, under more favorable conditions. The costs of failure are prohibitive when the success of an attack depends upon surprise and the defender is now on alert. Then, the "next" attack will be harder to engineer than the last. The costs of failure are high, too, when a current operation will expend offensive assets that are hard to acquire and replace. Trained commercial pilots with US visas that were critical to the September 11 operation were a significant (perhaps, irreplaceable) asset that al-Qaeda was apparently unwilling to invest without strong assurance that the mission would succeed. The attacker's aversion to these costs makes a defensive, denial-based strategy a viable *dissuasive* response to a potential attack.

US strategy can benefit greatly, then, by addressing the current threat in the broad and integrated terms of a deterrence problem. Principles of deterrence—vital to Cold War planning—can provide an effective framework for understanding and confronting the challenges posed by nonstate adversaries that can launch attacks on the US or its allies. A deterrence framework is useful even if there is no chance that a stable deterrence relationship can emerge between the US and some terror groups. Indeed, acknowledging the threat from al-Qaeda and other terror groups as a deterrence problem can facilitate an understanding of when, how, and why deterrence will fail. A government that cannot defend against all potential threats must appreciate how government strategies will affect the attack problem, reducing— yet also increasing—the attacker's incentives to strike.

Thus, this chapter discusses the political and military options for an attacker and a defender when viewed from a deterrence perspective.[2] It reveals that deterrence depends upon the attacker's ability to achieve its political purposes and, if it cannot, whether it can compensate by: (a) improving its attack capability; (b) increasing the payoff from an attack; (c) redirecting the attack to other targets; or (d) increasing the number of attacks. It concludes that the attacker's options are considerably constrained and, therefore, that deterrence opportunities are available to governments. Governments can capitalize on these opportunities by rejecting the notion of the risk-taking terrorist—out to destroy at all cost—in favor of the Cold War assumption of adversaries that establish standards for success, evaluate capabilities accordingly, and make reasoned choices about when, what, and where to strike.

Terrorist capabilities to attack a target

Although some terrorists seek targets of opportunity—a suicide bomber can wander in search of a crowd—even suicide bombers have a general target in mind and a plan for approximately what, where, and how the target will be struck. Therefore, it makes sense to evaluate how terrorists assess the adequacy of their capability against specific targets and strengthen that capability if necessary—as al-Qaeda did in plotting the attack on the World Trade Center. When the towers proved impervious to destruction from below (a bomb in the garage in a 1993 attack masterminded by later cohorts of bin Laden), it was targeted from above (by plane). Al-Qaeda capitalized on the peculiar vulnerabilities of the target (i.e. attacking the buildings at heights that caused them to crumble and trap a large number of occupants) and the full destructive potential of an aircraft (e.g. planes were fueled for transcontinental flights and had a small number of passengers to control).

In evaluating a target-centered strategy, however, the rational attacker must ask, "how much capability is enough?" Just as bin Laden did not send the nineteen hijackers of the September 11 attacks on nineteen separate hijacking missions, the rational attacker will settle for realizing less than the full destructive potential of its attack force to ensure that some necessary level of destruction is met, with some level of certainty. If so, the attacker must balance its goals and capabilities: specifically, it must devise a standard for success, determine the capabilities needed to meet that standard, adjust the standard if it taxes or underutilizes available capabilities, and determine the amount of (reserve) capability needed to protect against mission failure.[3] These considerations were crucial to US thinking about AD. Despite its assured "destruction" rhetoric, the US intended to spare most of the Soviet population in a retaliatory attack given, among other considerations, the declining marginal utility of further increasing the inflicted destruction. At the same time, the US sought to render destruction with "assuredness" and thereby sought a robust retaliatory force that could survive a worst-case Soviet attack (see Ball 1980: 168–77). That terrorists embrace this (conservative) decisional logic is revealed in evidence that Khalid Sheikh Mohammed, a chief architect of the September 11 attacks, had initially planned to hijack ten aircraft and attack targets on the West and East coasts of the US but was overruled by al-Qaeda leaders who were concerned (in the words of the *9/11 Commission Report*, p. 154) about the ambitious "scale and complexity" of the attacks; and in evidence that bin Laden does not appear to have required "full success" to realize his objectives. Although bin Laden had personally picked the target for the Washington-bound airliner that crashed into a field in rural Pennsylvania, he still rejoiced in the two hits on the twin towers of the World Trade Center and the direct hit on the Pentagon.

In determining how much capability to build into a plan to achieve critical goals and to protect against mission failure, the attacker must recognize that capability improvements do not come cheaply. In part, these costs reflect claims that secrecy imposes upon the terror organization: "to maintain the minimum level of anonymity needed to stay in the game, they must build in a wide range of fire walls, procedural constraints, and organizational redundancies into their day-to-day operations to keep their opponents guessing" (Hoffman and McCormick 2004: 245).[4] Even capable terrorist organizations must thus sacrifice some ability to adapt quickly to circumstance, to capitalize on fleeting opportunities, and to make efficient use of available resources. High costs also arise as consequences of capability improvements. Indeed, the "scale and complexity" of a terror organization or attack plan can produce confounding effects that compromise an operation.[5] This is illustrated by a multiple-attack strategy in which an attacker strikes more than one target to ensure that some number is destroyed ("hedge" targeting) or strikes one target multiple times to ensure that it is destroyed ("redundant" targeting).

The ability to coordinate multiple attacks, an al-Qaeda trademark, permits the attacker to increase the impact of an attack, to devastating effect. The large death toll on September 11 can be attributed to the attackers' ability to strike both towers of the World Trade Center in two separate attacks when attacks on two separate days would have been difficult. (That timing was imperative, here, is apparent from the fate of the last of the hijackers' Washington-bound flights: the last attack was doomed when the passengers, knowing about the prior attacks, attempted to take over the aircraft.) If the attacks are *independent* events (organized and executed apart from one another and maybe without knowledge of one another), multiple attacks are recommended because many attempts to destroy targets are more likely to succeed than fewer attempts are. If the attacks are *interdependent* events, multiple attacks offer the benefits of an "organized" attack. These include synergies that result from a division-of-labor among participants or from tactical deception. For example, one attack can set a trap for a second to spring. Illustrating this is the common practice in which one suicide bomber detonates their explosives to cause crowds to run toward a second bomber. This tactic results in a higher death toll than if the two bombers acted independently.

Still, the multiple-attack strategy comes with risks. Long recognized in organizational theory is that complex systems (or plans) increase the chances of unanticipated malfunctions as one part unexpectedly affects the performance of another part, which affects the performance of still others, and so forth (Sagan 1995). Even the redundancy built into a system to safe-guard against its failure can lead, through a surprising chain of events, to a catastrophic breakdown of the system. For example, military forces intended to backup other forces have caused "friendly fire" incidents in which troops accidentally fire on their comrades in the heat of battle. Or else, a system fails when some of its parts do not perform as expected, in the manner in which missing US air cover for (CIA-trained and equipped) Cuban nationals helped doom the Bay of Pigs invasion of Cuba during the Kennedy administration. The multiple-attack strategy is problematic due also to problems of scale. Adding people, technology, and resources to an operation makes it more "visible" and more likely to be sabotaged, then, through premature disclosure. That a multi-faceted, large-scale operation might be hard to hide is amply demonstrated by the hints left by plotters of the September 11 operation that could have divulged its existence. The hijackers carried passports that had been fraudulently altered and made statements on visa applications that could have been proven false; some of the participants were arrested or denied entry into the US before the attack; some of the hijackers had been trailed by the CIA abroad and into the US; Islamic radicals were known by the FBI to be receiving commercial pilot training in the US; some of the hijackers undergoing flight training were conspicuous due to their English language deficiencies, limited prior training and experience, and resistance to instruction; and many of the hijackers were actually subject to additional screening at airport check-points on September 11 (some having been flagged by computer as security risks). The probability of premature disclosure only increases if each attacker's behavior or knowledge (as revealed through interrogation) can compromise the entire mission. The irony of a multiple-attack strategy is that it can increase the chances that the attack will fail.

How much a future, large-scale, coordinated terrorist operation will invite problems of interdependence is hard to say. Within the al-Qaeda "organization," Osama bin Laden is said to have operated like a proficient

CEO of a modern corporation by adopting an "organizational framework and strategy incorporating multiple levels and both top-down and bottom-up approaches" (B. Hoffman 2003). Al-Qaeda has eschewed the hierarchic design of some terror organizations in favor of a compartmental structure of loosely connected secretive cells that coordinate their activities only as needed (for purposes of synchronization, intelligence, and so forth). The cells receive direction from the central al-Qaeda leadership (bin Laden) but operate with great autonomy (Shultz and Vogt 2003: 11). In practical terms, this means that bin Laden inspired and directed some operations—most notably, the "spectaculars," high-profile strikes on the US Trade Center, the US embassies in Africa, and the USS *Cole*. It means, too, that al-Qaeda was a tangential participant in many plots to which it is credited. Al-Qaeda has accepted ideas and financed promising projects from local organizations and employed operatives who can select targets, and obtain the financing, logistical support, and personnel, for an operation. This was true of the failed plot to bomb Los Angeles International Airport (B. Hoffman 2003). Al-Qaeda has also forged alliances with other radical groups in the pursuit of shared objectives. The merging of forces with Abu Musab al-Zarqawi's network in Iraq is the most visible—and perhaps important—of these contemporary alignments. Indeed, with attacks on its central leadership, al-Qaeda has increased its reliance upon local organizations to plan, initiate, and execute attacks. But even these smaller-scale operations involve coordination among multiple participants and links to external groups. Thus, the effects of failure could be far-reaching. Because terrorists operate as parts of larger networks,

> disruption can sow suspicion and distrust within a terrorist organization far beyond the cell that is disrupted, and materials that are confiscated when a cell is broken up often provide intelligence leading to the disruption of other cells—even cells in other countries.
>
> (Pillar 2004: 118)

For those who seek to disrupt the functioning of these organizations, their vertical and/or horizontal structure creates challenges and opportunities. Complex organizations are often surprisingly robust in their ability to tolerate multiple failures or attacks on their components. Networks that are fully connected—with *some* nodes that have large numbers of linkages to other nodes—are highly tolerant of failure when the nodes are subject to random disruption (Albert *et al.* 2000: 378–82). Then, the system can degrade gracefully—remaining intact and performing longer—because it is more likely (given simple probability laws) that the smaller numbers of less essential (less-connected) nodes will be destroyed than more-essential ones. The system will continue to perform as long as critical nodes are intact. Of course, attacks on terrorist "systems" (networks) might not be random, and the key nodes can be singled out for destruction. Even "random" attacks could quickly disable an interconnected system in which the disabling of some parts (the "key nodes") causes others to fail, or a differentiated system in which parts (e.g. individuals) perform specialized tasks and the loss of a part produces a loss in critical capability.[6]

Terror planners are aware of this when designing their "systems," that is, their organizations and attack plans. The key nodes in an operation— the mission organizers and commanders—can try to protect themselves (and their operations) by limiting how much the participants know about an operation, by staying aloof from the operatives (e.g. contacting them anonymously via the Internet or controlling when, where, and how contacts occur), and by keeping operatives (and attack teams) apart. In fact, all these techniques were employed by the planners of the September 11 mission. Still, the capture of some of the September 11 participants could have led to the capture of others. Mohamed Atta, the lead hijacker, was linked to a hijacker who was on a terrorist "watch list" when using the same contact address to book his flight; another hijacker shared the same telephone number with Atta; still another shared a post office box with the hijacker on the watch list (Gellman *et al.* 2006). Hypothetically at least, capturing participants could have reduced the size of attacks teams below critical levels, compromised key participants (the pilots) needed for the operation, or jeopardized the entire operation if resulting disclosures (of plans to hijack US aircraft) put the US on alert.[7]

Whenever the attacker chooses to increase its capability, then, it must be wary of unintended consequences that reduce the probability of mission success. The intensity and pervasiveness of these effects, which only multiply under conditions of extreme secrecy, give the attacker considerable reason for caution.

The shift toward smaller-scale, locally-based terror operations in Europe (and elsewhere), which draw from a large disenfranchised Muslim population, could ease the logistical challenges of a terror attack. But it will not end them; and these challenges are likely to remain great in big operations against distant US targets. The reality is that the "attack network" is vulnerable compared to the vast and amorphous terror networks that the US has targeted with its offensive counter-terror strategy. This is because terrorists must increase their profile when on the offensive. When conducting operations in foreign lands, terrorists must emerge from the shadows, coordinate their efforts, and engage in novel activities. These yield obstacles to a terror organization, many unforseen, that increase its chances of exposure.

Terrorist payoffs from an attack

Another rational strategy available to the attacker to offset the possibility that an attack will fail is to boost the value of an attack. In other words, one answer to a risky operation is to make it worth the risk. This can be accomplished, for example, by increasing the explosive power (or shrapnel production) of a bomb or engineering an attack on a physical structure to inflict higher damage or casualties. A useful by-product of this strategy is increasing the chances that the mission will succeed. For example, the ability of a single atomic bomb to produce widespread destruction at a national port of entry allows an attacker to overcome challenges that could compromise the success of an operation. The attacker need not worry about detection by cargo inspectors on arrival, arranging transportation or escaping detection when moving the bomb to, into, or around a city, and storing the bomb within the target country for eventual use.[8] But the value of an attack is not easily manipulated, for a number of reasons.

First, terrorists have almost always opted for conventional weapons attacks that are limited, at the high end, in the amount of damage they can inflict. Although the lethality of terrorist attacks increased somewhat in the 1980s and 1990s, the number of people killed in terrorist attacks around the world has been relatively small, the September 11 attacks aside. In the last century, "no more than 14 terrorist operations killed more than 100 persons at any one time" and "no single operation had ever killed more than 500 persons at one time." Overall, in the almost three dozen years leading up to the 2001 World Trade Center attack, around 1,000 Americans had been killed in domestic and international terrorist attacks (Hoffman 2002: 304).[9] The reality is that physics narrow the range over which conventional explosives will inflict their effects and, conversely, that scientific principles can be harnessed by the defender in building construction and barriers and checkpoints to keep bombs at distances to mitigate these effects.[10] In consequence, the trend has not been toward ever more destructive terror methods producing ever more deadly effects. Recent evidence actually suggests the opposite. The techniques in recent high-profile attacks were fairly simple—in the Madrid attacks, thirteen backpacks and packages were left on trains and then activated using the alarms on cellular phones. The London suicide-bomb attackers used fewer and lighter-weight bombs made from commonly available materials.

Second, nonconventional weapons—with their plausibly high death tolls—might not be the wave of the future. Terror groups have not found governments willing to hand over nonconventional weapons technology; nor have these groups had much success in developing these technologies independently. Despite popular claims that anyone can assemble a nuclear, chemical, or biological weapon from information available in public sources (e.g. the Internet), biological and chemical weapons have rarely been employed in terror attacks; and, when used, they have produced but a limited return on the resources invested given formidable problems of development and weaponization, that is, achieving necessary levels of concentration and dispersal. The notorious Aum Shinrikyo religious cult, after many failed (and unnoticed) attempts, killed only a dozen people in its Tokyo subway attack in 1995. With assets of a billion dollars, hundreds of scientists in their employ, and millions of dollars invested in a sophisticated infrastructure for chemical and biological weapons research and development, the cult could do no better than putting sarin gas in plastic bags and then piercing them with umbrellas. The track record of Islamic fundamentalists is similar in this regard: the attempt to magnify the effects of the 1993 World Trade Center bombing by packing hydrogen cyanide around the explosives failed when the chemicals vaporized in the bombing. Although practice could lead to improvement, history does not suggest that groups that have used nonconventional weapons are peculiarly adept at producing larger numbers of fatalities (La Free *et al.* 2005).

Even terrorists that overcome the sizeable barriers to development and weaponization must still make the hard choice between tactics that boost the *assuredness* of destruction and those that maximize *destruction*. Releasing organisms or poisons in a confined and controlled space, such as a building, greatly enhances the effectiveness of an attack but also increases the number of spaces that must be assaulted (or the number of post-infection contagious contacts that must occur) to produce large-scale death; in turn, contaminating a wide area risks a diffuse and ineffective result (Gurr and Cole 2000: 41–79; Koblentz 2003: 105–9). Consequently, the dissemination of enormous quantities of a toxin or chemical might be needed to produce a small number of fatalities.

Apart from problems of acquisition, terror organizations might not seize opportunities to acquire new weapons. For a terror group, the advantages of acquiring new weaponry must be weighed against the disadvantages of introducing a weapon that could bring undesired attention to an operation, is hard to develop in secrecy (e.g. a nuclear weapon) or to integrate into an organization with a compartmental structure, is insufficiently labor-intensive to motivate personnel through their involvement in attack preparations, or could effectively disable the organization when tested or handled incorrectly (principal victims of terrorist action are the terrorists themselves). (On these liabilities, see Jackson 2001.) The advantages of acquiring new technology must also be weighed against the advantages of retaining and acquiring weapons with which the users are familiar and adept. The use of the suicide belt—how to wear it, how to pack the shrapnel, and how to calculate the effects of the explosion—is a science in the hands of skilled practitioners of Iraq and the Israeli-occupied territories. If anything, success comes from a capability to improvise on the margin. The deadly effectiveness of guerilla and terror campaigns owes to the comfort and the adaptability with which operatives use established technologies. It should be remembered, for instance, that the unprecedented destruction of the September 11 attacks ensued from a creative application of a low-tech approach: attackers, armed with box cutters, hijacked planes and crashed them into buildings. The use of "improvised explosive devices" by Iraqi insurgents illustrates the art of building on current approaches. The insurgents have capitalized on peculiar vulnerabilities of targets (e.g. the unshielded underside of US Humvees), paired new and old technologies (e.g. lasers were introduced as triggering devices when garage door openers and cell phones were subject to radio frequency jamming), used variants of existing methods (e.g. larger bombs and shaped charges to penetrate armor), and adjusted tactics for increased effect (e.g. planting roadside bombs in inauspicious material such as garbage and planting follow-up bombs or positioning snipers to kill bomb-clearing personnel) (see Anderson *et al.* 2005). For that matter, terror groups have a strong incentive to play on their comparative advantages in nontechnological areas. Successful terror campaigns have relied, in general, upon an abundance of cheap resources, strategic advantages (such as proximity to targets and ethnic unobtrusiveness), and the skills and commitment of the participants. For instance, suicide bombers—because they do not need to plan an escape—can home in on targets and destroy them without delay or caution. The act of suicide also magnifies the political impact of an attack by signaling a high level of dedication and allows the perpetrator, in death, to serve the cause as a hero or "martyr" (Hoffman and McCormick 2004: 248–50).[11]

If terrorists were open to new technologies and could employ them effectively, chemical or biological attacks might not be useful for selling a message to a broad audience. True, there is considerable prestige available to an organization that employs a weapon that has been the sole prerogative of states and can produce violence on an unprecedented scale. After all, the impact of the atomic bomb dropped on Hiroshima in 1945 stemmed from the "exceptional" status of the weapon—the toll inflicted by a single bomb in a single strike.[12] The simple detonation of a "dirty bomb," which spreads radioactive materials through a conventional explosion, could have a highly disruptive effect on an economy, forcing the shut-down of contaminated buildings and entire neighborhoods over long time periods. But the use of nonconventional weapons could be counterproductive: the effects of a biological or chemical attack are hard to contain and can evoke compassion for the visibly suffering victims from an audience that might otherwise be receptive to the terrorist message (due, in part, to normative prohibitions on poisoning that are present in many societies).[13] In this regard, Post (2005: 149–50) notes that "social-revolutionary," "nationalist-separatist," and "single-issue" terrorists generally shy from acts that would alienate their popular support base by killing indiscriminately. What is significant then is how few global conflicts have led to indiscriminate terror attacks on civilians: mass-casualty attacks have been highly focused geographically—in places such as Iraq, India, Russia, and Sri Lanka—and only a small number of groups are responsible for a large percentage of the attacks (Asal and Blum 2005: 153–5).

Certainly, mass-casualty attacks came into their own in the last decade, for which al-Qaeda deserves considerable credit. Indeed, al-Qaeda has actively sought to obtain WMD—bin Laden has said so in his public communications (Frost 2005: 55)—and will likely try to use them if acquired. Still, the *priority* given by al-Qaeda to procuring these weapons could be reduced by the realization that conventional weapons have properties unmatched in biological and chemical attacks. Put indelicately, big explosions make for "great television." Horrific fireballs consuming critical targets (and giving the victims some anonymity) provide terrorists with payoffs that cannot be found in slow death—perhaps weeks, months, or years after an attack.[14] Not surprisingly, then, the insurgents in Iraq have widely distributed videotapes of their suicide bomb attacks through television networks and the Internet. They have also adopted techniques that permit maximum media exposure. It seems no accident that the attacks on the World Trade Center, the Madrid train system, and the London subway all occurred in the morning. Although the last two attacks were obviously meant to coincide with the rush hour (to increase the death toll), morning attacks have heightened visibility when images of the wreckage and the dead and injured are replayed throughout the day in Europe and the Middle East and then, again, when the news reaches a US target audience as it awakens within hours of the attack.

Third, the value of a target might be fixed or insensitive to the actual number of people killed or levels of physical damage inflicted. Thus, quite apart from whether terrorists can produce widespread destruction is whether they want to do so. As Brian Jenkins (1975: 15) famously observed, "terrorists want a lot of people watching and a lot of people listening, and not a lot of people dead." Who was killed, no less than how many are dead, is important to explaining the impact of an event, judging from the momentous effects of some historical assassinations (e.g. the Archduke Ferdinand). Moreover, the psychological and political impact of a violent act can be disproportionate to the human toll. Bloody airport attacks by Palestinian groups (and their cohorts) in the early 1970s, which did not leave large numbers of bodies (by contemporary standards), could make anyone, anywhere, feel vulnerable to attack; and, in the long-term, it does not appear that an association with persistent, high-profile—but generally small-scale—terrorist acts has hurt the visibility of the Palestinian cause. For that matter, the paradox of the dramatic, costly event—the bombing of Pearl Harbor or the sinking of the *Titanic*—is that it is seen as a "one time only" event and then but a historical event without meaning as tragedy. The sinking of the *Titanic*, a tragedy of monumental proportions in its time, is the inspiration behind popular slot machines operating after September 11 in casinos in Atlantic City, New Jersey—a short drive from New York City—and Halifax, Nova Scotia—the final resting place for many of the tragedy's victims.[15] Big operations such as the September 11 attack can also "raise the bar" for future operations. Anything less than an operation of its magnitude might be taken by the target audience as a "failure" of sorts—a sign that the perpetrators are losing strength. This perception can take hold because big operations (by virtue of their costs and risks) are hard to deliver with sufficient frequency to sustain political momentum. In part, for these reasons, al-Qaeda appears to be reducing its emphasis on *truly* spectacular events.

This is not to say that al-Qaeda is impressed by the philosophy that "small is better." On this point, some researchers have responded to the Jenkins' quote by noting that terrorists want a lot of people watching *and* a lot of people dead (Simon and Benjamin 2000); and it could as easily be said that terrorists want a lot of people watching by leaving a lot of people dead. Indeed, the phenomenal death toll on September 11, at least as much as the powerful visuals of hijacked airliners exploding into buildings, explains the intense public reaction to events of that day; and the multiplication of casualties has been a consistent priority of al-Qaeda—communicated in its manuals, strategy sessions, pronouncements, and actions. Thus, the quote does not accurately describe the "new terrorist" with which the US must now contend.

But this is not necessarily bad news for the intended targets. That terror organizations might pursue ambitious and deadly attacks could help the defender given the logistic problems of extravagant attacks. Even operations with low death tolls present complications for the attacker which cannot be overcome simply by increasing the damage from an attack. When terrorists appreciate this, deterrence is in play at some level.

Terrorist selection of targets to attack

When encountering obstacles that reduce the likelihood that a target can be struck with success, the attacker has another option: it can redirect its attack toward more vulnerable (less protected) targets. Indeed, it is in their choice of targets, not weapons, that contemporary terrorist organizations have sought to accomplish their political purposes (Asal and Blum 2005: 155).[16]

A wealth of available attack choices permit the attacker to strike not just unusually vulnerable or valuable targets but especially vulnerable *and* valuable ones. For example, the exclusive screening of carry-on luggage at airports would allow terrorists to smuggle bombs onto planes in the cargo hold. Given a choice between hitting two targets of identical value—say, one that the attacker has a ninety percent chance of destroying in a strike and another that the attacker has a ten percent chance of destroying in a strike—the rational attacker will choose to attack the more poorly defended target. The quandary for the defender is knowing what to protect, then, when it cannot protect everything. More specifically, the question is what to select for protection and what to exclude from protection given two complications.

First, lesser defensive priorities are still inviting targets. In fact, it is hard to conceive of a target that offers at least some political dividends to a terrorist if attacked successfully that is not also valued by the targeted government or society. Thus, from the defender's perspective, everything appears vulnerable. For instance, attacks on any of a large number of buses and trains are useful to an attacker that seeks to convey that everyone, anywhere, at any time, is susceptible to attack—that people risk their lives doing the ordinary. Not only could such attacks have extensive social and economic repercussions, when raising doubt about the security of one or all modes of public transportation, they are difficult to stop. Bus and train stations are designed for public access and efficiency, not for security. Even isolated attacks on anonymous individuals can have enormous terror-producing effects. The Washington, DC sniper attacks in the fall of 2002 show convincingly that a small number of small-scale attacks can instill fear in a populous, affecting business, schools, and social life. Despite the thousands of law enforcement personnel assigned to the case, an elevated police presence throughout the region, and an anxious and attentive community willing to aid the police effort, a lone sniper team managed to hold the US capital and its suburbs hostage over a periods of weeks through a series of random killings. The defender simply did not know when and where the next attack would occur and could not prepare accordingly. As always, the problem for the defender is that it can try to protect what it values most but this is wasted effort if the attacker's interests center on another target, that the defender also values.

Second, the attacker can choose to attack less-protected targets when defensive commitments create offensive opportunities. By defending certain targets at the expense of others, the defender redirects the terror threat toward "softer" targets. It was this paradox that motivated the CIA to dispatch counter-terror experts to small towns throughout the US in the summer of 2004 to advise them that they faced an increased terror threat with the heavy security planned for the political party conventions in Boston and New York (Johnston and Jehl 2004). This target shift was also evidenced in Iraq when insurgents moved from attacking less vulnerable US targets (e.g. bases, convoys, and targets in Baghdad's walled green zone) to Iraqi civilians, recruiting centers, police stations, diplomats, government officials, and Shiite mosques. It is evidenced, too, in al-Qaeda's shift to soft targets in Europe, Asia, Africa, and the Middle East.

When evaluating its capabilities and setting its priorities, then, the defender encounters the long-recognized dilemma of a state that seeks to deter attacks on its interests. By designating areas of the world that are in the national interest to defend, a state implicitly excludes other portions of the world from protection and invites attacks on those lesser interests. Conversely, by claiming lesser interests as "vital," a state risks depreciating the credibility of its promises to defend any and all of these vital interests if attacked. The credibility of the defender's claims are at issue in part because of what can be described as a "commitment" problem. The defender can "signal" its terrorist adversaries that it will remain vigilant against any and all terror attacks, as the US has done by establishing the Department of Homeland Security and adopting various security reforms—but ambitious commitments invite challenges. The more the defender commits to defend, the more likely it is that the challenger will test the defender's resolve and/or capabilities (all other things being equal).

Illustrating these dynamics are the Israeli government's "no negotiation" policy toward terrorists. The logic of the policy is that conceding to terrorist demands rewards terrorism and only encourages future attacks. When Palestinian terrorists took hundreds of Israeli school children as hostages, however, in 1974 at Maalot—foretelling an incident three decades later in which Chechens took a considerably larger number of school children as hostages in Beslan—the Israeli Knesset met in emergency session and effectively withdrew the "no-negotiation" pledge. Whether or not the Israeli government was seeking only to buy time for a rescue effort, an offer to negotiate was a tacit admission that the costs of maintaining the pledge were prohibitively high.[17] Such unsustainable positions risk eroding the credibility of all government commitments.

Even if the attacker believes that the defender has the *intention* to defend its interests—most certainly true of the defense of targets on national soil—the defender's credibility is in question when the attacker doubts the defender's *capability* to respond successfully to an attack. Because the defending government cannot do everything and be everywhere at once, the government cannot devote resources to the protection of targets in proportion to their value. The sheer number of places where large numbers of people congregate—among them, the most vulnerable and sympathetic portions of the population (e.g. school children)—make a universal defense strategy impossible, and dangerous. By attempting to defend more than it can, the defender risks undermining its ability to deter attacks.

Yet the options for the attacker, too, are constrained given its desire to strike targets of value to a government or society. This limits the range of available choices to targets that governments have a strong interest in defending. Symmetries in value between the attacker and defender are arguably more the rule than the exception when terrorists desire to hit societies "where they hurt" and to magnify the political, social, and/or economic effects of an attack. Reflecting upon the events of September 11, bin Laden reportedly hailed the effects on the airline industry and US economy of an attack "lasting one hour only" (Flynn 2004a)—an attack that cost al-Qaeda half-a-million dollars to execute cost the US economy half-a-trillion dollars (Mueller 2005: 40). By destroying a few airliners on that day, bin Laden brought US airline traffic to a halt. Years would pass before airline travel returned to its pre-attack levels and then only with a significant diversion of national resources into airline security. High and growing costs were also inflicted in the London subway bombings. Weeks after the bombings, billions of dollars in revenue were estimated to have been lost to the London economy, security costs had greatly increased, and the number of passengers remained significantly depressed (Jordan 2005). Indeed, the evidence is that the impact of so-called spectaculars reverberates beyond the limited threat those attacks present—that people tend to exaggerate the chances that they, too, will be victimized. Because of these very consequences, governments can focus their defenses on what terrorists want to attack. For example, these symmetries arguably abet the US strategy of protecting nuclear power plants more than oil refineries, airline transportation more than bus stations, and the US capital more than other US cities (e.g. the now-rescinded, thirty-minute passenger-mobility restriction on flights in and out of Washington National Airport). It also played to the US strategy, immediately after September 11, of concentrating protective resources upon the disarming of airline passengers—preventing them from boarding planes with potential weapons—over screening stowed aircraft luggage for explosives. Assuming that terrorists had less interest in killing hundreds of people when, by hijacking an aircraft and using it as a weapon, they can kill hundreds (perhaps, thousands) of people, destroy a physical structure, and receive credit for pulling off *another* September 11 style attack, the US could focus upon what it regarded as a costly attack scenario. Such symmetries also explain the current US focus on screening luggage rather than cargo shipped on passenger planes. Without insider industry knowledge, terrorists run a high risk that a shipped device will explode onboard a cargo aircraft, a presumably low-value target, rather than a passenger plane (Elias 2003).

Although the defender might still be unable to offer a robust defense of the numerous targets that terrorists want to strike, the defender can benefit, as well, from important capability and informational advantages (asymmetries). These limit what the attacker can gain from an attack and/or force the attacker to accept risks and costs in planning and executing an attack. These advantages can boost the credibility of a defender that can capitalize on a variety of (denial-based) strategies.

First, the defender can rely upon a *limited defense*. A limited defense is meant to concede ground—given the prohibitive human and financial costs of a robust defense—and only contain the damage that is suffered in an attack. Viewed from a deterrence standpoint, a limited defense attempts to reduce the value of an attack. The current enforcement of flight restrictions around Washington, DC illustrates this. With the high costs of around-the-clock patrols by military aircraft, an identification zone and narrower flight-restricted zone within which aircraft are closely monitored concentrically encircle the US capital. Although an array of military capability—from F-15 and F-16 aircraft to ground forces equipped with surface-to-air missiles—can quickly assemble to bring down a suspicious civilian aircraft when the order is given, it is unlikely that the implementation sequence will unfold within the time period available to shoot down an approaching aircraft or that the order will be given barring extraordinary evidence of hostile intent.[18] Because officials thus concede that the system is intended to stop a *second attack*, not a first (Hsu 2004),[19] this limited defense can be said to aid deterrence. The system reduces the value of an attack by thwarting the catastrophic outcome (i.e. multiple planes crashing into multiple buildings) that motivates the attacker.

Second, the defender can resort to a *partial defense* to boost the costs and risks of an attack. For instance, by reducing key vulnerabilities in high-value targets, the defender can deprive the attacker of easy victories and force it to adopt more expensive and dangerous tactics to accomplish its objectives. Simply reinforcing and locking the door to the airplane cockpit, and keeping the door locked under all circumstances, dramatically increases the challenge for an attacker that seeks to gain control of a passenger aircraft. Indeed, the defender can adopt single measures that reduce the value of an attack *and* increase the costs to an attacker. Protecting critical nodes (e.g. in the electric power grid) or choke points (e.g. railway or highway tunnels) that could produce highly disruptive effects if attacked reduces the payoff from attacking those sites (by offering a limited defense) and forces the attacker to adopt cost-ineffective methods to achieve attack objectives (e.g. attacking electric transmission towers and lines rather than substations). The same principles work in the defense of symbolic targets. New measures to protect the Washington monument in Washington, DC consist of unobtrusive retractable posts and knee-high level, interlocking granite walls to protect against a large, explosive-carrying vehicle (Dvorak 2005). An implicit assumption of this defense is that full protection is not required given the challenges and cost of alternative means of attack. For instance, an aerial attack would inflict damage that was disproportionately low relative to the price and risk of the operation.

Third, the defender can utilize a *flexible defense* by allocating resources as needed to blunt an anticipated or actual attack. Consequently, local defenses can be designed only to limit damage from an attack through a "holding action" until reinforcements arrive in the form of light, mobile quick-response military forces, special weapons and tactical (SWAT) teams within police departments, or emergency responders such as hazardous materials teams, medical personnel, fire departments, and help arriving from other municipalities, states, or countries. Flexible defense is also embodied in organizational and/or technological systems that permit a coordinated and informed response to an attack. With respect to the latter, a system is being deployed in the Washington, DC subway that joins chemical sniffers, close-circuit television cameras, and computers that can model and track the spread of a toxin in the event of a chemical attack. Among its advantages, then, the system is said to reduce emergency response times, to permit the informed control of devices through which toxins are spread (such as vents that open onto streets and moving subway cars that act as "plungers" in spreading toxins through subway tunnels, exits, and vents), and to identify safe areas where subway passengers can be channeled and/or rescued (I. Hoffman 2003).

Fourth, the defender can resort to a *selective defense* in which resources are allocated to combat more damaging threats rather than less damaging ones. For instance, security is often disproportionately tight at sports events that attract tens of thousands of people. Smaller groupings of people are attractive targets, but the possibility that terrorists could kill and injure a large number of people in a single incident—in a celebrated venue— makes these events important to defend. Likewise, security can focus on more- over less-damaging modes of attack. For example, US officials can worry more about a destructive car and truck-bomb threat to homeland civilians and structures than the threat from dismounted suicide bombers. Notwithstanding the high "cost-per-casualty ratio" achieved by organizations such as Hamas with their suicide-bombing campaigns (Dolnik and Bhattacharjee 2002: 113), the effectiveness of these attacks is lost over long physical distances. This is due to the costs of relocating the attack force, problems keeping operatives committed and focused in a distant land, and the small pool of motivated attackers who can blend into and function in the targeted society. In contrast, Europe—with its open borders, large Middle Eastern populations, and proximity to North Africa—has proven more vulnerable to "small-scale" terrorist attacks than the US has.

Fifth, the defender can engage in *forward defense* through actions that affect the attacker's basic capability to carry out an attack. Embracing the philosophy that "the best defense is a good offense," "defense" can motivate an offensive military strategy. The intent, then, is to weaken the opponent, keep it off balance, and force it to divert resources that could be used in an attack toward ensuring its own survival. The liabilities of military preemption are considerable (as discussed in Chapter 5). But an offensive strategy also subsumes less provocative measures that include multilateral efforts to strengthen global monitoring and enforcement mechanisms that impede the transfer of chemical, biological, and nuclear weapons technology. As importantly, they subsume domestic efforts aimed at depriving terrorists of financial and public support, and monitoring and penetrating terror cells.

Sixth, the defender can engage in *defensive screening* efforts in which populations are "filtered" as they pass through key access points. The intent is to locate those who fit a suspect demographic or behavioral profile and to subject them to additional screening (i.e. searching and/or questioning). Although it is common for critics of demographic "profiling" to claim that terrorists can always change their profile—for instance, by employing Norwegian instead of Middle Eastern or South Asian suicide bombers—the reality is that Scandinavian suicide bombers are in short supply and the defender can prepare with that in mind. Profiling understandably provokes concerns of resulting discrimination against a particular racial, ethnic, age, or religious group; and even profiling based on behavior can lead back to demographic considerations. Actions such as withdrawing large sums of money from the bank, purchasing one-way tickets, or even walking through a certain neighborhood appear suspicious when performed by some people and, innocent, by others. Profiling can also be ineffective. Profiling will not work if the size of the suspect population is too large to target for rigorous screening, government agencies focus on physical characteristics at the expense of useful behavioral markers, and members of a "suspect population" retain useful information out of fear that they will inadvertently incriminate themselves, family members, or friends, or damage their community. In principle, though, screening underlies all efforts to monitor the flow of people or goods at some distance from possible targets. For instance, a "profiling" of sorts underlies the US monitoring of container shipments into the US. US customs inspectors focus their scrutiny on "untrusted" shipments from problematic areas of the world and/or that involve importers that have not built a record for clearing customs (Flynn 2004b: 90). No more than five percent of cargo containers are actually opened by US customs agents (Blustein and Pincus 2006).

Seventh, the defender can construct a *triggered defense*. The defender need not stay on maximum alert all of the time but can husband resources and go on alert when a possible threat is identified. Once mobilized, the defender is better able to combat the threat and to appreciate its actual dimensions. Signs of a prison break can trigger a general "lock-down" to counter both the immediate threat (e.g. prisoners who are trying to escape) and unknown other—perhaps, bigger—threats to which the precipitating incident is linked. Similar triggering occurs when a security violation in some US airports leads to an order to "dump the concourse" which requires the re-screening of all airline passengers (Flynn 2004b: 77). Triggering is also involved in decisions to shut-down a subway system when some trains are attacked, to heighten security precautions for all forms of public transportation when one mode is attacked, and to heighten security in US public transportation when the subway system in another country (e.g. Britain) is hit. On a grander scale, triggering is involved in the US color-coded alert system in which the government indicates, and changes its procedures with, the perceived level of the domestic terror threat and the defense condition (DefCon) system that places US military forces at a lower or higher state of readiness. Of course, for a triggered defense to work effectively, the defender must be able to separate credible warnings from false alarms. Otherwise, repeated false alarms could incur significant costs, reduce readiness by inducing general complacency (the "cry wolf" syndrome), and allow the attacker to observe and surmount defensive preparations and procedures. In fact, a "false alarm" could then be triggered by terrorists as a diversion.

Eighth, the defender can engage in a *random defense*. Just as bargaining theories of deterrence relied upon a "threat that left something to chance" (Schelling 1960), the defender can choose to protect some targets or take some actions periodically and/or unpredictably to increase the risk to the attacker. The logic of risk manipulation supported the random screening of US airline passengers in the aftermath of September 11. Random screening—let alone the screening of but one-in-ten passengers—appears to make little sense from a defensive perspective. It does make sense from a deterrence perspective (especially if it is assumed that the ten percent chance of being screened combines with other uncertainties with which the attacker must contend). The ten percent detection probability is that much more effective as a deterrent if attackers in a group (the nineteen hijackers) each have a one-in-ten chance of being screened and the detecting and detaining of any one attacker might impair or compromise a terror operation. Then, random screening serves a selective defense that focuses on combating a (September 11 style) multiple-attack scenario.[20] From this perspective, it is logical that New York officials responded to the second set of London subway bombings with a random screening of packages and backpacks on the New York subway and that US officials board ships and inspect cargo randomly though some illicit cargo goes undetected.[21]

Ninth, the defender can employ the *spatial defenses* that were used in the Cold War era to strengthen nuclear deterrence. One aspect of these defenses is *mobility*. Just as the nuclear powers relied upon mobile submarines and land-based missiles to keep nuclear forces secure from attack, governments can harness mobility to protect government leaders from assassination. The schedule and movements of the US president are often kept secret and, in times of emergency (e.g. September 11), the president can remain mobile (e.g. Air Force One) or be taken to an undisclosed, fortified location for protection. Another aspect of these defenses is *dispersion* to reduce the value of a *target* (in contrast, a limited defense reduces the value of an *attack*). Just as the nuclear powers chose not to co-locate all of their nuclear resources—creating an inviting target for attack—the US President and Vice President currently avoid attending the same public events. Dispersal is the logic behind recommendations that dangerous chemical facilities not be allowed to operate near population centers or that trains carrying dangerous cargos be routed away from urban areas: hazardous material are less inviting targets when distanced from potential victims (see, e.g. Flynn 2004b: 121).[22]

Finally, the defender can rely upon *defensive uncertainty*. Despite al-Qaeda's legendary ability to obtain information on targets (through open sources and active surveillance), not all of the strengths and vulnerabilities of a target will be known to an attacker. Available floor plans and maps (such as the one that led to the accidental US bombing of the Chinese embassy in the former Yugoslavia) might be incorrect, dated, or lacking fine details (e.g. the location and capabilities of an alarm system). Or else, uncertainty could arise from unresolved engineering or practical issues that are implicit in an attack scenario. Illustrating this is the considerable controversy about whether the reactor core of a nuclear power plant could survive a direct hit from an aircraft (Walsh 2003: 183). Uncertainty could also result from a deliberate policy of defensive concealment. For example, security units seek to multiply their effectiveness by reducing the predictability of their patrol schedules and staffing and by withholding information about their counter-terror tactics and procedures. Inevitably, all partial and flexible defenses have some amount of useful uncertainty built into their performance, unintentionally or by design. A visible airport security presence—though for passenger screening—offers some protection against any and all attacks on airline transportation because the defender *could* stumble onto an attack.

Thus, deterrence can be strengthened when the defender creates favorable capability and informational asymmetries using a number of defensive approaches, alone or in combination. Indeed, deterrence effects could multiply enormously through a packaging of approaches—as illustrated by security at a hypothetical gathering attended by a government leader. The leader can be protected through a selective defense in the form of bodyguards, a full screening of people in close contact with the leader, a partial defense (against certain kinds of attacks) through screening with metal detectors or explosive-sniffing dogs, random screening of all people in attendance, and uncertainty about where the leader will be sitting and how and when the leader will be entering and departing the venue. In combination, these imperfect approaches offer the deterrent advantages of a *layered defense*. The variety of possible combinations of defensive approaches is too large to discuss in full. Suffice it to say that which combinations are usefully employed depends on available defensive resources, the nature and intensity of the threat, and the defender's tolerance for error. Random screening makes little sense if an assassin is known to be in a crowd just as random screening or selective defense of urban targets make little sense if terrorists are known to possess a nuclear weapon that can devastate an entire city. (Some guidelines for employing these approaches are presented, however, in the final section of this chapter.)

It is easy, then, to understate the range of options available to the defender and to overstate the options available to terrorists. But terrorist groups have not taken full advantage of opportunities to exercise "high-end" options (i.e. spectacularly destructive attacks) or even "low-end" ones that are well within these groups' capability and promise a considerable return. If terrorists have rejected either set of options based on their cost or value, the capabilities of the attacker are inadequate given its objectives, and deterrence is again in effect, at some level. Policymakers can magnify these effects with appropriate defensive strategies.

Terrorist selection of the number of targets to attack

Given limited payoffs from attacking single targets, the attacker can choose to attack multiple targets. A multiple-attack, multiple-target strategy (henceforth, "multiple-attack") is desirable, in part, because of its psychological advantages. A diffuse attack can instill fear in the target audience by making the attacker appear able and willing to strike anywhere, anytime, at will.[23] But the multiple-attack strategy also presents problems for the attacker.

The advantages and disadvantages of attacking in numbers are apparent through the use of the modeling techniques employed in Chapter 4 to assess the strengths and limits of missile defense. They show that there are hidden costs and potentially limited benefits in a multiple-attack strategy. (The figures for these analyses are found in the Appendix on p. 183 and are discussed in more detail in the chapter notes.) The defender's position is viewed more favorably here than in Chapter 4 for three reasons. First, defending certain high-value targets against terror attacks is arguably less challenging than defending the US homeland against missile attacks. Much depends upon the number and variety of terror targets and their relative value to the attacker and defender. Second, the defender can tolerate some level of failure in protecting against conventional terror attacks. Whereas one nuclear bomb launched by a rogue state against one city would be devastating, a defense can be said to work against conventional terror attacks if reducing casualties and property damage to a level that the defender can tolerate. Third, the defender has reason to believe that a conventionally armed terror group will withhold an attack if necessary to avoid high costs or to manage risk. A rogue state that, with premeditation, would launch nuclear-armed missiles against the US is discounting of the future (i.e. reckless and/or suicidal) and a poor candidate for deterrence through punishment or denial.

To appreciate the deterrence challenges and opportunities in defending against multiple terror attacks, it is useful to start with a simple observation: the laws of probability guarantee, other things being equal, that the probability that attacks will meet their objective increases with: (a) the actual number of attacks, and (b) reductions in the minimum number of attacks that the attacker wants to succeed. The chance of destroying at least two targets is greater with six or eight than with four attacks; the chance of destroying at least two targets is greater than the chance of destroying at least four or six targets in eight attacks; and the chance that all attacks will succeed is greater with two or four than with eight attacks (see Figure A.1).[24]

It should be recognized, however, that the chances that a mission will succeed—as defined by the desired minimum number of successes for a given number of attacks—decline disproportionately with the effectiveness of the defense and the ambitiousness of the attacker. Against a capable defense, the probability that the attacker will reach its mission goal is low for all but the attacker that defines operational success at levels of destruction (e.g. two successes out of eight attacks) well below that which could be inflicted by its forces in the best-case scenario (e.g. eight successes out of eight attacks).[25] In other words, a tough defense imposes potentially prohibitive costs on all but those willing to expend considerable resources to achieve relatively small gains. Even a moderate defense (say, one that can foil fifty percent of attacks) could impose prohibitive costs on some attackers (e.g. a fifty percent chance that two of four attacks will fail).[26] Thus, the rational attacker has a strong incentive to avoid well-defended targets unless their destruction offers exceptional benefits.[27] These conclusions apply even to an attacker that is prone to take risks.[28]

Although high levels of target protection can admittedly be hard to obtain, there are reasons for the defender to be optimistic about its chances of success. First, an attacker might try to make its offensive count by seeking a large number of successful attacks on a variety of targets—the scenario that the defender is best positioned to deter. Even attacks on soft targets could yield unfavorable results for an attacker that aspires to levels of success beyond those achieved on September 11 (i.e. three successes out of four attacks). Second, as the ambitiousness of the attacker grows— specifically, as the attacker seeks a larger number of successful attacks *and* a high rate of success—the marginal utility of building on weaker defenses grows too. For instance, the payoff for five successes out of five attacks must be over thirty times greater, in the attacker's view, to justify attacks on individual targets that can be protected successfully seventy-five percent of the time over targets that can be protected over fifty percent of the time, and almost 100 times greater to justify attacking individual targets that can be protected successfully ninety percent of the time over those that can be protected seventy-five percent of the time (see Figure A.2).[29] Even slight improvements in a weak defense can dramatically reduce the attack incentive for an attacker that is loath to accept failure: for an attacker seeking eight successes out of eight attacks, upgrading a defense so that it can stop but one-in-four attacks can reduce the attacker's chances of eight successes to around one-in-ten.[30] The defender also reaps large marginal returns in building on a weak defense against smaller-scale attacks. For instance, improving from no defense to an ability to stop one-in-four attacks produces a formidable defensive result—a bit more than a one-in-four chance that all four attacks will succeed. Although the defense is extremely challenged against an attacker that will settle for a low rate of success (e.g. a one-in-ten rate of success), the defender can significantly improve its capability to deter an attack with incremental improvements in capability from a wide range of starting points.[31] This bodes well for the utility of a number of defensive approaches—defensive uncertainty, random defense, and flexible defense—that, by themselves, might not elevate defensive capabilities to the high levels required to deter an attack. These defenses could inexpensively supplement an existing defense and dramatically boost the defender's capability to deter multiple attacks.

Of course, this assumes that the attacker strikes *different* targets, when a conservative attacker encountering tough defenses could well favor redundant targeting. But hedging offers the attacker strong advantages over redundancy. First, redundant targeting undervalues the potential of the strike force. Although redundant targeting is a viable strategy for assuring the destruction of targets that hold exceptionally high value to the attacker (all other things being equal, two *independent* attacks on a valuable target are more likely to destroy it than a single attack is), such targeting does not offer the possible returns from attacking multiple targets (of identical value). The reason is that an attacker must concede much in exchange for the assuredness of destruction. Put simply, by launching two attacks on one target, the attacker effectively surrenders its ability to destroy two different targets, that is, to realize the full destructiveness of its attack force.[32] For example, evidence from Palestinian suicide bombings indicates that using two attackers (over one) in a single operation increased the death toll but not to levels obtained when attackers engaged in separate operations (Harrison 2006: 191). Second, redundancy (more so than hedging) invites the multiplication of risk through the contagion of failure.[33] The fact that multiple individuals or teams are engaged in a given attack makes it more likely to be compromised through premature disclosure, mutual interference, or the failure of one team to perform an essential task. True, attack plans can survive the loss of an attack team, the defender is often unable to put together available clues in time to stop further attacks, and captured participants might not know nor reveal details that could compromise other parts of an operation. Still, even small failure contagion rates can have a decisive impact on the overall success of a mission. This can occur when the number of interdependent attacks is large, a growing number of compromised attacks can compromise still others, and the probability that an attack will fail due to the contagion of failure combines with the effects of other factors that reduce an attack's success.

Admittedly, terrorist leaders are unlikely probability theorists, and they might well exaggerate their ability to carry off an attack. After all, terror groups exist often by virtue of a megalomaniacal belief about the likely impact of a terror campaign; and, regardless, the chances of success cannot be known with certainty. But terrorist groups can still appreciate unfavorable statistical realities that emerge through intelligence gathering, operational setbacks, arrests or near-arrests of operants, and outright failures of prior attacks. As the attacker is best positioned to know, plans fall apart, bombs explode prematurely, and attacks fall short of their objectives. By forcing the attacker to accept a large "margin of error" in assessing risk, these possibilities give added weight to the most important conclusions, here, that (a) relatively weak defenses can be employed credibly to deter the most dangerous attacks, and that (b) moderate defenses can effectively deter somewhat less dangerous ones. Hence, partial, selective, or random defense approaches (among others) are successful deterrents if they help boost the attack failure probability to these required levels. Limited and flexible defense approaches are similarly effective when the attacker seeks a large number of successes, both in absolute terms and relative to the number of attacks.[34]

Conclusions and implications

Because terrorists do not set themselves up for failure, governments are positioned to deter attacks. Deterrence is possible because the preferences of governments and terrorists overlap, improvements in terrorist capability can yield insufficient returns in attacks, and capability or informational asymmetries complicate the devising and organizing of attacks. Then, the attacker has a formidable task: it must methodically plan missions to avoid compromising effects, cope with uncertainty about the likely outcome of an attack, mobilize resources for conducting those operations, avoid defeats that hurt the group's ability to perform, and attack targets that governments have reason to protect. As a result, the attacker has plenty of reasons to delay or withhold an attack, which could advantage the defender. The attacker might choose not to attack, might choose to attack a less consequential target, or might delay an attack giving the defender time to pick up signs that an attack is forthcoming and to prepare accordingly. *All three outcomes are desirable from a deterrence standpoint.*

A deterrence strategy might not stop all attacks; in fact, it might not stop attacks in which relatively simple devices are used to kill large numbers of people. Unfortunately, too, deterrence could cause the attacker to change targets and/or methods resulting in a more damaging attack than the one that had been deterred. But the defense-based deterrence strategy is based on the assumption that, over the long term, the defender is better off with the strategy than without it or when relying upon an offensive-based strategy. The principles underlying the strategy are straightforward: "instead of trying to protect every conceivable target against every imaginable form of attack" (Jenkins 2002: 29), the defender seeks to cause the attacker to accept greater costs and risks and/or a reduced prospect of gain in planning and executing an attack. When governments accept these principles, a number of useful defense policy guidelines emerge.

First, governments must safeguard their priorities by protecting against *possible* worst-case attacks on national citizens—their lives, livelihood, and property—and *probable* attacks that reflect the goals of the attacker (e.g. the World Bank buildings or Wall Street). In selecting their protective priorities, governments must distinguish the practical value of a target from its patriotic or sentimental value, likely effects from less-likely ones, and short-term consequences of an attack from long-term effects. Thus, governments might recognize that certain structures require intensive protection (e.g. national monuments) only when lives are at stake, that the economic value of a target is less sensitive to its replacement costs than to human costs that produce long-term economic effects,[35] and that even severe vulnerabilities (e.g. to a nuclear attack) might not justify correctives. This is because governments can viably choose to defend probable targets before improbable ones, remedy vulnerabilities that are inexpensively fixed over those that are costly to repair, or solve problems that respond poorly to post-hoc remedies rather than those that respond well. Therefore, a viable strategy could center on protecting transportation links—and airlines in particular given their value and vulnerability—and places in which large numbers of people congregate.

Second, governments can protect targets by conceding their vulnerability. Because effective defenses are possible that only increase the *chances* that an attack will fail—governments have at their disposal useful (random, flexible, and partial defense) approaches that reduce the value and increase the costs and risks of attacks on targets. For example, the reinforcing of cockpit doors goes a long way toward fixing the aircraft-takeover problem and allows the transfer of protective resources from passenger screening to aircraft protection. It is unnecessary to expend precious resources in confiscating scissors or other sharp objects from passengers when the main danger is now an explosion on board an aircraft. A passenger that gets stabbed on an aircraft is no different than a passenger that gets stabbed on a Greyhound bus. As long as the cockpit doors remain locked, the pilot is dedicated to landing the aircraft, and an aircraft is not catastrophically damaged, most airline passengers are likely to survive a foreseeable onboard incident.[36] Thus, governments must draw a distinction between tolerable and intolerable levels of destruction. In all likelihood, people will die and/or property damage will occur in a terrorist attack regardless of what governments and private interests do to prevent it. The operative question must be how best to expend resources to contain and otherwise limit the consequences of an attack.

Third, governments can protect targets by not increasing their vulnerability. Through regulatory efforts aimed at hardening, zoning, transporting, or policing, governments must defend or isolate facilities and vehicles that when destroyed can have devastating collateral effects. Chemical plants and trucks and trains carrying hazardous materials are inviting targets when proximate to urban areas, and efforts should be devoted to "distancing" these potential weapons from lucrative value targets.

Fourth, governments can protect targets even when doing so incompletely. Defensive measures succeed, controlling for cost, if reducing casualties and damage below levels that would otherwise have occurred. During the Palestinian suicide bombing campaign, bombers that were halted at the first lines of defense—a checkpoint or street—were likely to do less damage than when attacking stores, bus stops, buses, or cafes (listed in ascending order of value, based on casualties inflicted in an unchallenged attack).[37] (See Harrison 2006: 197–9.)

Fifth, governments must respond, but not over-respond, to threats. Flexible defenses rely upon scarce and depletable resources, and the danger should flexible defenses become static or overused is that they will be unavailable for other contingencies. For instance, officials must carefully select days on which to shift to higher alert, for calling alerts too frequently makes them a pointless exercise ("another false alarm"). When officials call alerts around holidays (e.g. the Fourth of July) that have "special significance" in the US, when the attacker attaches no special meaning to those occasions or believes that unpredictable attacks have added potency, the defender has left itself more vulnerable, at least by some small measure to attack on regular days by effectively having wasted "alertness."

Sixth, governments must control national gateways and key corridors of attack. By screening traffic through immigration offices and airports, governments can increase the overall risk to a terror operation that involves large numbers of personnel. Governments must also act to impose risks late in the attack plan. It is one thing to deny entry into a country of an operative whose papers are not in order; it is quite another to nab an attacker at the preparation stage (when guns or explosives are being purchased or specialized training is being sought) or (at a checkpoint) during the execution of an attack. Attackers should not be permitted to assume all risks up front when the security of a terror operation is least likely to be compromised (inasmuch as operatives are not carrying incriminating documents, communication devices, materials, or weapons that could signal an impending attack) and the penalties for the participants are relatively mild (e.g. deportation rather than death in a failed attack).

Seventh, government must adopt covert measures signaled overtly. Covert surveillance can trap a suspect, but a security presence can deter an attack only when advertised. In the manner in which highway signs effectively warn drivers that speed limits are being enforced with detective radar, visible devices and procedures that allude, somehow, to the existence of hidden ones are a useful deterrent. These visible defenses need not be *transparent* defenses: signaling the full capabilities of a defense is advisable only when defenses are impenetrable. The best deterrent might be an occasionally strong, albeit somewhat unpredictable, security presence.

Eighth, governments must anticipate new threats but can focus on established ones. Because terrorists are innovative and can use surprise to advantage, governments should attend to possible changes in threat. In studying intelligence information, Internet chat rooms, radical websites, and reports of break-ins, suspicious activities, or weapons thefts, governments must recognize signs that terrorists are eyeing new targets, adopting new weapons, or altering their methods of operation. At the same time, governments must appreciate that terrorists play to their own strengths and modus operandi and copy what worked in the past. Imagining what terrorists *could do* is a useful exercise; but terrorists might have neither the desire nor capability to do what we fear most. This means that governments can capitalize on the learning curve, as the adversary's behavior becomes more predictable with time. Ultimately, Israel did become more proficient at defending aircraft and then buses from attack when each in turn became the centerpiece of a Palestinian terror campaign. Thus, there is no substitute for heightened and directed vigilance by law enforcement personnel and the public around the kinds of targets that were hit in the past. Useful then is a situational awareness that flags as dangerous a package that is left unattended on a bus or train, and municipal efforts to create a clean, open, and well-lit environments (on subway platforms) in which an unattended package will stand out. Also useful is attending to familiar warning signs. For instance, airline screeners can increase their effectiveness by focusing on suspect behaviors (e.g. the purchase of one-way tickets, last-minute ticket purchases, and bulky and inappropriate clothing that could conceal explosives). Less useful is raising the national alert level and encouraging people to buy duct tape and plastic to seal their home windows in the event of a biological or chemical weapons attack. Although the cost of focusing on the past is unpreparedness for the future, allowing the terrorist some initial success is the inevitable price of vigilance.

Ninth, governments must anticipate attacks but should recognize the value of belated action. Governments are best equipped to discover terror networks after the fact—that is, post-attack—when resources can focus on the persons responsible (before the trail has gone cold). Past terror operations have left "fingerprints" at the scene that can be traced to the culprits. For instance, shortly after the Madrid attacks, the police discovered detonators and cassette recordings of the Koran in a van that had supposedly been used by the bombers; later, a ringing cell phone rigged to a bomb alerted police to an unexploded device with a defective trigger at a train station (Sciolino and Van Natta 2005). It is much easier to follow trails from clues—explosive fragments, rental cars, unexploded devices, documents, or the bodies of suicide attackers—left at the scene of an attack and to use that information to arrest participants and to thwart other attacks than it is to pick up signals of an attack in a noisy environment and to follow the trail back to the conspirators. The aggressive and successful pursuit of those responsible can have the deterrent effect of requiring that terrorists expend additional effort to cover their tracks.[38]

Tenth, governments must give priority in defensive planning to conventional threats but must work to prevent terrorists from developing or acquiring WMD. Given their potential destructiveness, nuclear weapons in particular allow the attacker to "change the game." The low-profile, high transportability, and immense destructiveness of a nuclear bomb *at its time and place of arrival* within a target country mean that it would be very difficult to detect and stop a nuclear attack without advance intelligence of an impending attack. The threat of a nuclear attack—either with a nuclear bomb or spread of radioactivity with a conventional device—is real and worrisome. By one estimate, 20 tons of HEU exists at 130 often poorly guarded civilian research facilities in forty countries around the world. The breakup of the Soviet Union alone supposedly left enough HEU and plutonium within former Soviet territories to construct 60,000 nuclear weapons.[39] Moreover, the Russian nuclear weapons inventory consists of thousands of nuclear weapons with varying dimensions, applications, mobilities, and yields. These weapons could pose a major threat to the world's cities given their huge numbers, the declining state of Russian nuclear security, and the possibility that these weapons could be stolen or sold (some of these weapons have allegedly "disappeared").[40]

The possibility of a nuclear attack admittedly bolsters the argument for an offensive-based denial approach to countering terror groups and their state supporters. It should be said, though, that denial-based policies also include stiffening controls over weapon stockpiles and hazardous materials, cooperative strengthening of multilateral non-proliferation regimes, reinforcing the threat to punish terror organizations that possess these materials and states that supply them, and focusing intelligence efforts to obtain warning of preparations for such an attack. Governments would do well, though, *not* to plan disproportionately for a homeland interception of a WMD attack. The challenges of defense are too formidable to tolerate a significant diversion of resources to that aspect of the attack problem. (For a contrary view, see Flynn 2004b.) Governments must recognize that conventional weapons pose the far more probable threat and a threat against which governments are best able to defend.

Finally, governments must seize and resist offensive opportunities. Killing or capturing terrorist leaders abroad and destroying their bases of operation can weaken terrorists in the short term; but these measures can also swell the attacker's ranks and capabilities and create a less complex, less visible, and highly motivated terror adversary. Recent operations in London, Madrid, and elsewhere suggest that aggressive, high-profile military actions can create an insidious terror problem in the form of a less hierarchic, locally based adversary. To the extent that potential operatives fit into indigenous communities, know their ways around potential targets, and need not rely upon outside logistic support or direction, it becomes easier for terrorists to pursue their goals without detection. This inconspicuous, competent, and self-reliant adversary could be more difficult to deter.

Some of these suggestions seem intuitive and there is evidence that the US government has become more sophisticated in its defensive approach to the terror problem. For instance, the US Transportation Security Administration that manages the US air security system has self-consciously struggled with the trade-offs forced by limited budgetary resources and has rescinded some of the security measures that were in place after September 11 (Goo 2005; US GAO 2006) and added others (in the wake of the disruption of planned 2006 attacks on US-bound flights from London). But a deterrence strategy requires more than piecemeal and sporadic efforts to address the security problem. First and foremost, it requires a balancing of priorities within and across sectors, domestically and internationally, with sensitivity to the evolving threat. How resources are spent and which efforts are emphasized must reflect US comparative advantages, areas of vulnerability, US priorities, and the dangers at hand.

7

FINAL THOUGHTS

The strengths and limits of deterrence

The September 11 attacks shook the US and dramatically changed its position in the world. Within a few years of the attacks, the US government had reorganized its executive branch, passed (and bypassed) legislation to affect the relationship between the government and its citizens, reinterpreted long-standing international conventions to justify harsh treatment of foreign detainees, adopted a decidedly unilateralist approach to foreign policy, fought wars in two countries, and then moved to rebuild them through violent, expensive, and painstakingly slow efforts. At the end of the day, the nagging question is whether the US is more secure for its far-flung and costly policies. Current strategies beg for a change in course.

Main conclusions

If it is assumed that rogue states and terrorists will not simply flail themselves against the US or its allies at the first opportunity—indifferent to the magnitude of damage inflicted, the value of targets hit, and the risks and opportunity costs incurred—the logic of deterrence must join the pure logics of offense and defense in US strategic planning. The challenge for US policymakers is in recognizing the strengths *and* the limits of offenses and defenses and, thus, how each might reinforce or undermine deterrence relationships between the US and its various adversaries. Because policymakers have not yet risen to the challenge, critical deficiencies plague all three "pillars" of contemporary US policy—preemption, missile defense, and homeland security. These deficiencies are as follows.

First, by pushing the advantages of an offensive-denial strategy of preemption to combat rogue states, policymakers overstate the value of military action against well-protected nuclear arsenals and understate the benefits—indeed, the inevitability—of US reliance upon a coercive strategy. There is little to suggest that rogue-state leaders are inclined to commit the ultimate act of suicide terrorism (the "mother of all suicide terrorist acts")— launching a nuclear attack on another nuclear-armed state. The logical predicament and behavioral record of these states suggests, in fact, that they can be deterred. For that matter, the logical predicament for the US is that it might rely upon a punishment strategy, after a preemptive US strike, to obtain access to nuclear sites to ensure that they do not pose a threat.

Second, in pursuing a missile defense, the US pairs a hypothetical offensive problem with an expensive, deficient, and dangerous solution. It fails to appreciate the "boomerang effect" if US adversaries counter a defense by weakening controls over nuclear weapons, by utilizing the bargaining leverage that US defenses afford, by channeling the arms competition into new weapons or attack modes, by seeking missiles in greater numbers, or by increasing the threat to US allies.

Third, in addressing the problem of terrorism, the US exaggerates the effectiveness of an offensive-denial strategy of preemption. Killing or capturing terrorist leaders abroad and destroying terror training bases can help the US obtain valuable information from captured operatives and documents, and grants the US the initiative to confront terror groups before they can carry out their attacks. But effectively attacking terror groups is complicated by the problems of targeting the terrorist infrastructure, by the supply of new leaders to replace assassinated ones, by the decentralization, secrecy, and networking of terror groups that results when they are attacked, and by the damaging consequences of a *high-profile* US military presence in many parts of the world. Indeed, "taking the fight to the enemy" through preemptive military action—comes with onerous costs. As the US intervention in Iraq demonstrates, actions that inflame passions, feed an existing sense of grievance, and fail to provide stability and security can swell the attacker's ranks and capabilities without improving the defender's ability to fight terror. Such actions can actually drain resources from that fight. The military price tag for the Iraq and Afghanistan operations is predicted by authoritative government sources to run well over 800 billion dollars, exceeding the costs of the Vietnam war by hundreds of billions of (constant) dollars (Weisman 2006).

Fourth, in overstating the value of an offensive approach, the US understates the contribution of defending against the terror threat. Because terrorists do not relish failure and must plan missions to avoid compromising effects, to manage uncertainty, to mobilize resources to conduct those operations, and to avoid costly defeats, dutiful US attention to homeland security could give terrorists abundant reasons to forestall action or withhold an attack. Such a delay or change in plans could work to the defender's advantage.

An effective deterrence strategy is not predicated, then, on a US willingness to make threats, adopt bold policies and audacious tactics, or assume commitments that incur high costs. Neither is it predicated on rigid assumptions about the overall superiority of the offense relative to the defense, or punishment relative to denial, as a basis for strategy. Each has its place, and deficiencies, under some conditions. What is needed is not a grand solution but a judicious mix of strategies.

In this spirit, Figure 7.1 juxtaposes the offensive/defensive and punishment/denial principles that inform current US policies toward rogue states and terrorists and an alternative juxtaposition that draws from the arguments presented in this book. Clear from Figure 7.1a is that current US policies are equally balanced between offense and defense but are decidedly biased toward denial principles. The US has acted on these denial principles with large-scale, offensive, counter-terror efforts in Afghanistan and Iraq, lower-profile actions spread throughout the world, and a considerable bureaucratic and financial investment in homeland defense. The US has acted on denial principles, too, by promoting and adopting preemptive tactics against rogue-state arsenals (e.g. in Iraq) while pursuing an ambitious NMD. Certainly, punishment has a role in current US policy. US policymakers have placed political, economic, and military pressure on North Korea, Iran, and Syria for their alleged pursuit of WMD and/or support for terror groups. For that matter, US policymakers insist that their denial-based policies in Iraq have the additional benefit of sending coercive messages to North Korea and Iran and attribute the new-found cooperativeness of Libya to its lessons learned from the US invasion of Iraq. But punishment is a subordinate feature of current US policy and reflects at best a grudging US acceptance of the limits of a denial strategy.

In contrast, Figure 7.1b indicates that US policymakers are correct to emphasize a denial-based counter-terror policy—but one with a defensive core. Offensive-denial and offensive-punishment based strategies can aid the defense but should not impinge upon it, as can happen when offenses draw resources from the defense or compound its challenges. Conversely, US deterrence strategies toward rogue states must favor a punishment-based approach. The US should make it clear to any and all US adversaries that the price would be steep for an unprovoked attack upon the US or its vital interests; and the US can adopt a variety of (unilateral and/or multilateral) offensive economic, political, and military (denial-oriented) measures to dissuade these and other states from developing their non-conventional arsenals. These include the discriminate uses of force against suspect cargo and limiting trade with potential weapons proliferators.

There is strong justification, of course, for believing that deterrence is but a subordinate—even unimportant—aspect of the relations between the US and its rogue-state adversaries. Rogue-state leaders—like all leaders—are susceptible under duress to common decision-making maladies—miscalculation, overconfidence, and misjudgment—that could send conflicts spiraling out of control. The irony of Saddam Hussein is that, in attempting to secure and extend his power base, he took actions that ultimately led to his demise. A lesson to be learned here is that rogue-state

(a) Current US policies

	Offense	Defense
Punishment	Rogue states Terrorism	
Denial	**ROGUE STATES** **TERRORISM**	**ROGUE STATES** **TERRORISM**

(b) Recommended US policies

	Offense	Defense
Punishment	**ROGUE STATES** Terrorism	
Denial	Rogue states Terrorism	Rogue states **TERRORISM**

Figure 7.1 Competing policy implications of a deterrence strategy

leaders will not always do the right thing when it matters most. But the danger in efforts to counter the uncharacteristic behavior of these states is in causing them to act uncharacteristically—to forgo caution and relax safeguards on the use of non-conventional weapons. In this important sense, the lesson of Saddam Hussein is that all states are capable of neglecting their interests and resorting to costly and self-defeating actions. The lingering US involvement in Iraq offers powerful testimony in this regard.

It is harder to argue that terror organizations, such as al-Qaeda, that seek to hurt the US will hesitate if given the chance. There are a number of reasons to suppose that al-Qaeda will attack the US *when it can*. First, al-Qaeda and its affiliates are implacably hostile to the US, grandiose in self-evaluation and purpose, and thereby relentless in their violent efforts to force the US to retreat from its global policies. Second, al-Qaeda, like other terrorist organizations, must consistently move forward to survive. To accept long periods of preparation or inactivity is to lose the focus, cohesion, support base, and momentum with which these organizations maintain their effectiveness. Third, a point of no return can eventually be reached when even a cautious attacker cannot be deterred. An attacker that is presented a one-time (fading) opportunity to assassinate a government leader might take the shot, though it is unlikely to succeed; and an attacker facing impending doom might strike out because it has everything to lose by waiting. It seems that bin Laden was anxious to strike the US in the months leading up to September 11 out of fear that the mission would be compromised by the large number of attackers operating within the US. But to say that an adversary is dedicated and hostile—even risk-prone—is not to say that it is irrational and therefore unwilling and unable to delay or discard attack plans that appear risky or costly. Apparently, al-Qaeda has abandoned plots that were fairly advanced in their development. These include the 1995 plot to bomb a dozen US commercial jets over the Pacific, halted after a bomb had been tested on a Philippine airliner.

Maintaining perspective

Viewing current US security challenges as "deterrence" problems fosters a supple and subtle understanding of the US security dilemma and the constraints on US policy options. It also presents a useful opportunity to learn from errors of the past. In this regard, the Vietnam war offers powerful lessons about the danger of missions that become their own reality. Millions died in that war, and through its collateral effects (Cambodia), based on false premises about the adversary and its goals, the viability of the US military strategy, the usefulness of indicators for judging military progress, and the severity of the threat. The underappreciated lesson of that war, which is now being relearned in Iraq, is that security challenges must be kept in perspective and policies must be adopted that are commensurate with the threat.

This applies certainly to US efforts to address the terror threat given that the average person, in their lifetime, runs a near-zero chance of being hurt or killed in a terrorist attack. Mueller (2005: 28) puts the terrorist problem in perspective when he observes that even counting the September 11 attacks:

> the number of Americans killed by international terrorism since the late 1960s . . . is about the same as the number killed over the same period by lightning—or by accident-causing deer or by severe allergic reaction to peanuts. In almost all years the total number of people worldwide who die at the hands of international terrorists is not much more than the number who drown in bathtubs in the United States.

Of course, bathtubs do not probe for points of weakness and increase their deadly efforts as opportunities permit. This is important if the terror threat is small *because* the US has protected itself. But the counterfactual reasoning that "things would have been worse had we not responded" can be used to validate any and all policies—no matter how extreme, dangerous, or costly—including the hundreds of billions of dollars that the US has already invested in Iraq.

In a sense, the problem is that governments and the public take the problem of terrorism too seriously. The admonition that the principle danger to the nation is "fear itself" clearly applies to the current threat, which works less directly through death and destruction and more through instilling anxiety or panic within a general population and a public willingness to accept security at any price. Failure in the war on terror is obviously measured by civilian dead and wounded yet it is also measured by self-inflicted wounds incurred when alerts are called based on specious information, civil liberties are too quickly sacrificed, legal principles are reinterpreted to aid and abet torture, resources are wasted because normal budgetary safeguards have been loosened and specious programs are funded in the name of "homeland security," troops are sent to engage in wars that produce new generations of terrorists, and the deficiencies of policies are belatedly recognized, at great cost in national blood and treasure.

Admittedly, the analysis has been neutral on a variety of pressing issues. These include the role that the military should play in domestic society, the aggressiveness with which governments should monitor their citizens and curtail their civil liberties, and the discrimination that attends a policy of profiling individuals by ethnicity, race, religion, or nationality. These are vital societal issues that cannot be addressed in the limited space here given the topic at hand. Suffice it to say that no anti-terror policy is unproblematic; and all such policies risk doing more harm than good. Governments cannot protect everyone from everything and terrorists will have numerous opportunities to create havoc within an open society. There will likely be deterrence failures—even spectacular ones. People who enjoy rights of free speech and assembly are easy targets of terrorist action. Societies can adopt extreme measures to defend against every and all terrorist threats and, in the end, might be no more secure. In the insatiable quest for absolute security, the question is what societies are willing to give up in return. Ultimately, keeping the terror problem in perspective—immunizing societies to panic and over-reaction—is the greatest contribution a government and public can make to a "denial" strategy.

APPENDIX

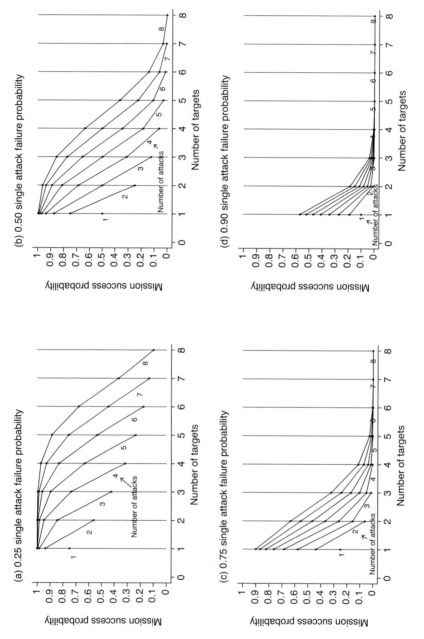

Figure A.1 Overall success with which targets are destroyed with hedging: four levels of target vulnerability

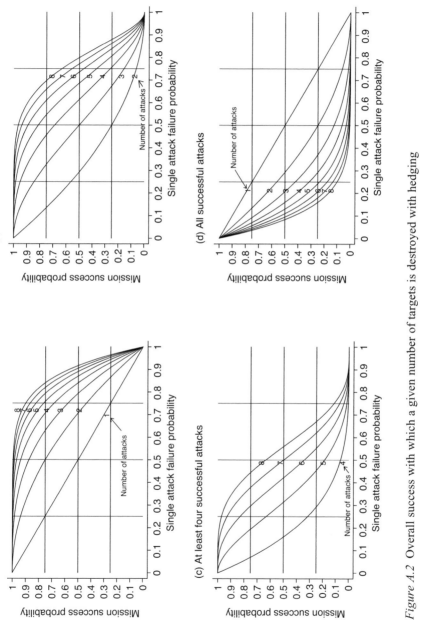

Figure A.2 Overall success with which a given number of targets is destroyed with hedging

NOTES

1 INTRODUCTION: DETERRENCE IN A CHANGING WORLD

1 The reference to "rogue" states in this analysis—a term that preceded the Bush administration (Hymans 2004)—should not be taken for acceptance of the many assumptions, including moral imperatives, that typically accompany this pejorative label. At a minimum, the term is useful shorthand for the multiple countries that the US identified as major policy concerns. Employing the term in the analysis also sensitizes the reader to the limits of labels when applied to any single country or set of countries. In contrast, the reference to "terrorist" groups in this analysis draws from the definition of "terrorism" that is used by the US State Department. It assumes that terrorism can be defined as a method of operation—specifically, acts of violence that deliberately target noncombatants for a political purpose (thereby intended to influence the behavior of governments or the public). The focus is on the behavior of these groups not their legitimacy or morality. The analysis does assume, however, that different terror groups pursue different goals and that these objectives must inform a counter-terror strategy. For instance, although governments can engage in acts of terrorism against their own people ("state terrorism"), these acts are not usually a direct threat to the security of outside countries.

2 Of course, Soviet missiles in Cuba carried some risk that it might assume control over these missiles or that threats against Cuba could trigger the use of the missiles.

3 Research suggests that the relative power and interests of the "initiator" and "defender" condition their behavior (Huth 1988; Huth and Russett 1988). Other research offers a psychological and political interpretation of deterrence successes and failures, e.g. Jervis 1989; Lebow and Stein 1989. On the "rational deterrence" debate, see Achen and Snidal 1989.

2 VIEWING THE PRESENT THROUGH THE PAST

1 I refer to these, elsewhere, as "political" war-fighters; see Lebovic 1990.

2 During the Cold War, when US nuclear strategists sought ways to preserve "deterrence" at some level with the Soviet Union, even into (nuclear) war, they were also addressing a compellence problem. Certainly, the US wanted to *deter* an all-out US–Soviet nuclear exchange that would incinerate both countries. Yet, there were compellence aspects to the problem, too. If the US had fired the first nuclear shot in retaliation for a Soviet conventional attack in Europe, the US

would presumably be seeking to *compel* the Soviet Union to renounce its territorial gains. If the Soviets had started the war, though, with the seizure of territory and full knowledge that the US would respond with force, the Soviets might have little desire to comply with US demands.

3 ON THE OFFENSIVE AGAINST ROGUE STATES: PUNISHMENT AND DENIAL STRATEGIES

1 The distinction between the two is confused when US policymakers refer to preemptive attacks when actually meaning preventative ones. I will refer only to "preemption" when the two options have similar implications.

2 That the image of the nonrational adversary lurks behind fears of the cost-indifferent and demonic adversary is conveyed when Payne, who decries the mischaracterizing of US policy as focusing on irrational adversaries, expresses concerns about "how the opponent interprets what is reasonable."

3 Perle served on the US Defense Policy Board and was widely regarded to be the intellectual leader of the so-called neo-conservatives within the Bush administration.

4 Thus, Iraq used illicit means and front companies to import technologies for *civilian* purposes to avoid bureaucratic entanglements and questioning when the technologies had military applications. See US Senate Select Committee on Intelligence (2004): 20–1.

5 This explanation receives substantial support from evidence found in the CIA report authored by Charles Duelfer (2004), former head of the Iraq Survey Group. In light of the report, it must be said that Hussein might also have been hiding his capabilities to *reconstitute* his weapons programs.

6 In this sense, the Iraqi strategy is arguably a rational variant of Israel's long-time nuclear strategy. Israel has chosen not to acknowledge possessing nuclear weapons when a public admission could provoke its neighbors and spur a nonconventional, regional arms race; yet the "open secret" that Israel possesses perhaps hundreds of such weapons—the implicit threat of disproportionate retaliation for an attack—could deter neighboring states from taking actions that threaten Israel's security.

7 US Commission on the Intelligence Capabilities of the United States Regarding Weapons of Mass Destruction (2005): 162.

8 See Howard (2004: 99–100); NTI (2006a). Inspectors were also concerned about the results of tests on Iranian centrifuges that indicated they had been used to enrich uranium. Iran's claims that they had been exposed to uranium prior to arriving in Iran appear to have been validated.

9 The relatively low priority placed on military threats and opportunities helps account for the willingness of Argentina and Brazil to eschew their nuclear weapons programs though the national security environment of these countries had changed little in the preceding years. It also helps explain the attractiveness of the nuclear option to broad domestic coalitions within some countries (in Iran, India, and Pakistan) and the willingness of some national leaders (in India and South Africa) to pursue these weapons with little military input in the development of the weapons or their strategies of use (on this, see Sagan 1996/97).

10 Nuclear weapons programs are of special concern to export-oriented business interests that oppose the inefficiencies of government-supported enterprises: "such programs often contribute to the ailments afflicting these countries' domestic political economy, such as the expansion of state power, the maintenance of

unproductive and inflation-producing military investments, and the perpetuation of rent-seeking patterns" (Solingen 1994: 139).

11 The North Koreans admitted to possessing a uranium program in late 2002 in meetings with a US delegation. See NTI 2006b.

12 Of course, the US is motivated to resist concessions that could have spillover effects in other parts of the world.

13 Of course, true to form, North Korea did not give away the store. The agreement said nothing about the timing of concessions—exactly when, and after what US actions, North Korea would have to open its doors to inspection and dismantle its nuclear programs.

14 Iran's president sent somewhat different signals (Moore 2006b). It appears as if Iran was using the Russian plan to head off Security Council action. At best, Iran saw a Russian-based program as a parallel effort that would help Iran acquire technological skills for a home-based nuclear enrichment program; and then Iran anticipated that it could eventually shift these foreign-based operations into Iran (Finn 2006b). The result was an unsuccessful end to the Russian–Iranian negotiations.

15 Counterfeiting activities account for a large percent of the country's export earnings (Kessler 2006b).

16 Preemption is one aspect of the Bush administration's broad-based policy approach, but one that the administration itself has chosen to emphasize (Levi and O'Hanlon 2005: 98).

17 For a good discussion of the technological challenges and unknowns, see Montgomery 2005.

18 The committee was chaired by Laurence H. Silberman and Charles S. Robb.

19 See US Senate Select Committee on Intelligence 2004: 29.

20 Civilian leaders can pressure the intelligence community to adjust its practices, for instance, through the power of appointment and by convening review panels (e.g. Team B assessments composed of outside experts) to scrutinize agency practices; and analysts are understandably reluctant to take contrarian stands when an administration voices strong preferences and there is a price to be paid, implicitly, for being wrong.

21 This pressure is that much more effective given a preexisting tendency within intelligence analysis to produce stronger conclusions than the intelligence information warrants. See Cordesman 2003: 429–31.

22 For an accessible discussion of the physical principles, see Glaser and Fetter 2005.

23 In fact, fearing the emerging taboo around the use of nuclear weapons, some Eisenhower administration officials apparently looked for opportunities to use these weapons (Freedman 2004: 70).

24 Consequently, revelations in the mid-1970s before a US Senate select subcommittee of CIA involvement in a number of high-profile assassination attempts worldwide (including numerous plots on the life of Cuba's Fidel Castro) led to an executive order prohibiting US participation in assassination plots. But this had the effect of creating new justifications and methods for targeted killings: the US hesitated when given an opportunity to kill Osama bin Laden in the late 1990s, whereas US forces in the course of military operations have targeted foreign leaders—such as Libya's Moammar Qaddafi in 1986 and Iraq's Saddam Hussein during the 1991 Gulf War. On the traditional norm against assassinations, see Thomas 2000.

25 Another includes establishing the pace of development and reform. The occupier must act quickly to capitalize on momentum and optimism after the defeat of a dreaded leadership and to avoid being seen as part of the problem rather than the

solution; at the same time, it must slow the transitional process to allow time to build efficient and accountable government institutions, to develop a functioning economy, and to heal societal wounds through the reconciliation of hostile elements within a country. On the trade-offs in the Iraq effort, see Henderson (2005).

26 Perito (2005: 12) notes, for instance, that the US experience with security in Operation Iraqi Freedom was "remarkably similar" to the US intervention in Panama of 1989.

27 Nations have cooperated, for instance, in joining the US-led 2003 Proliferation Security Initiative, which seeks a pooling of international resources for interdicting suspect air, land, or sea cargo to prevent the proliferation of WMD and their delivery systems.

28 For a pessimistic assessment of the performance of sanctions, see also Pape 1997.

29 The effectiveness of sanctions adopted against the Iranian regime could be compromised too by the renewal of interest in Iranian "regime change" (Baker and Kessler 2006).

30 On challenges of building a nuclear program through the "proliferation network," see Montgomery 2005.

31 Certainly, a plausible case can be made that North Korea violated the spirit rather than the letter of the agreement, which did not speak directly to uranium enrichment facilities. But the agreement did speak to North Korean obligations under the NPT and the 1992 denuclearization agreement with South Korea, which were inconsistent with a secretive enrichment program.

32 Such a reward strategy received the tacit endorsement of the Bush administration when it adopted a somewhat more conciliatory stance toward North Korea after trying a hard-line, non-accommodative approach.

33 This perception will predominate if the centers of regime control are targeted in the attack—an option currently under consideration against Iran. See Moore and Ricks (2006).

34 Koblentz (2003/4) argues that the variables and uncertainties of biological weapons *would* inhibit their effectiveness as weapons for deterrence.

35 Whether the targeted country would respond with nonconventional weapons is another matter. This led the US to a somewhat indeterminate stance during the 1991 Gulf War on whether nuclear weapons would be used against Iraq if it employed biological or chemical weapons. On this and the related "commitment trap," see Sagan (2000).

4 A DEFENSIVE-DENIAL STRATEGY AGAINST ROGUE STATES: NATIONAL MISSILE DEFENSE

1 NMD advocates do sometimes clarify the relationship between retaliatory deterrence and defense. On the "two sides of deterrence," see, e.g. Slocombe 1999. Their specific arguments suggest, nonetheless, that traditional deterrence thinking is inapplicable.

2 Whether a system is "limited" in architecture is in the eye of the beholder. Consequently, I use the term "cautious supporters" to distinguish those who propose that the US acquire the capability to defend itself from a handful of rogue-state missiles *and* strongly appreciate the physical limits of defensive technology and its confounding political and strategic effects. Although much of the current controversy centers on Bush administration efforts to construct a defense with a capability to attack warheads in their mid-course phase, as part of a layered defense that might also attack long-range missiles in their boost phase or

warheads at their terminal (atmospheric reentry) phase, even cautious supporters lend support to a small-scale mid-course defense (Daalder *et al.* 2000; Lindsay and O'Hanlon 2001). Unless stated otherwise, though, my arguments apply to all manner of NMD systems intended to defend against long-range missiles. (For an excellent discussion of these systems, see Lindsay and O'Hanlon 2001.)

3 AD advocates accepted most of these assumptions, in their broadest form, but believed them to have different implications than skeptics did. AD advocates generally agreed with (a), (d), and, perhaps, (c). (AD advocates could logically support defenses that protected the US second-strike capability.) On the other hand, AD advocates worried less about (b). In their view, a nuclear attack of any scale would likely trigger an all-out nuclear exchange (albeit some AD advocates sought limited nuclear options in case deterrence failed, blurring the distinction between AD advocates and their critics). For the most part, denial strategists accepted these assumptions but rejected the assumption (of retaliatory deterrence) that deterrence relies on punishment.

4 For their part, cautious NMD advocates articulate some of the worries of "new school" pessimists (Sagan, in Sagan and Waltz 1995) who see the new nuclear powers "as probable victims of misperceptions, technical mishaps and inadvertent actions they may neither fully anticipate nor be able to cope with" (Karl 1996/7: 94).

5 Cold War theorists acknowledged this interdependence; and, for instance, some AD advocates opposed missile defenses for their destabilizing offensive implications. On launch control and the potential for nuclear accidents, see Blair 1993; Sagan and Waltz 1995. On these problems as an effect of missile defense, see Glaser and Fetter 2001: 71–2.

6 On the Iranian case, see Cordesman 2004: 140.

7 Militaries have also been known to favor offensive strategies that promise quick ends to conflict and can be used to justify better weapons and larger budgets (see Sagan in Sagan and Waltz 1995).

8 These conditions are known as "negative" and "positive" control, respectively. See Seng 1997: 55.

9 The intelligence community has predicted that China will soon deploy mobile missiles that have yet to appear (Lieber and Press 2006). See US National Intelligence Council 1999.

10 Boost-phase defenses would destroy attacking missiles soon after they are launched when they are larger, slower, and more visible targets. The US could deploy boost-phase defenses (on land, in the air, or at sea) against North Korea without directly threatening China's entire nuclear force (Glaser and Fetter 2001: 52–4), but these defenses could still offer capability against missiles fired from some parts of China and require bases close to Chinese territory.

11 As one recent report put it, Russia's nuclear weapons may be the "greatest military danger" to the US "in consequence if not in probability" (CDI *et al.* 2001: 5).

12 On the considerable problems of targeting a country's mobile reserve force, however, even with nuclear weapons, see Glaser and Fetter 2005: 96–7.

13 Of course, the idea of a balance or imbalance is somewhat fuzzy. Reflecting their different intellectual commitments, Cold War-era thinkers engaged in fierce debates over whether one or both nuclear powers possessed usable advantages over the other. In the end, assessments of the nuclear balance had to address seemingly insolvable issues. Should the nuclear balance be measured before or after an attack? If the latter, after the initial attack or after a protracted nuclear exchange? Should measures of the balance center on the imposition of force or should coercive potentials of the opposing arsenals somehow play into the

analysis? By any reasonable interpretation of Cold War-era standards, however, rogue states are at a disadvantage.

14 For a good discussion of Schelling and recent research on the credibility of threats, see Danilovic 2001.

15 Strategists who were concerned about coercive advantages borrowed assumptions from AD and pure war-fighting doctrines.

16 For a more formal discussion of apparent resolve, bargaining, and missile defenses, see Powell 2003.

17 See Lindsay and O'Hanlon 2001: 63–4. Even the primitive nature of offensive weapons can be a countermeasure of sorts: their instability in flight and break up on reentry made Iraqi Scud missiles more challenging targets for Patriot missiles.

18 The Patriot tested perfectly but failed in the field (Lewis *et al.* 1999: 39).

19 More formally, the analysis is based on binomial probabilities—the likelihood that a given number of warheads will or will not survive the defense—where the number of warheads fired represent the original sample size (n), the survival ratio (1–the kill ratio) represents the selection probability (p), and the number of warheads that survive represents the number of successes (x). The likelihood that a given number of warheads will survive is thus understood as the probability that a certain number of cases will be selected (x) from a population of a given size (n), given the chance that any one case in that population will be selected (p). Probabilities are calculated from: $(n!/(x!(n–x)!))*(p^x(1–p)^{n–x})$. The analysis does not take account of missile accuracy and reliability problems that could prevent adversary missiles from accomplishing their objectives. The effects of these factors can be surmised by downwardly adjusting the number of attacking missiles or upwardly adjusting the defense kill ratio. On the accuracy of Iranian missiles, see, e.g. Cordesman 2004: 37–40.

20 If defenses short of ninety-five percent effectiveness fall within the ninety-five percent confidence interval, there is an even higher probability that a warhead will penetrate the defense.

21 Some knowledgeable Chinese civilian specialists on missile defense and China's nuclear arsenal have indicated to American experts that China believes it needs a 1:3 ratio of warheads to interceptors in order to ensure its minimum second strike capability. I am grateful to my colleague David Shambaugh for this information.

22 On a quantitative Chinese reaction to US missile defenses, see also CDI *et al.* 2001: 10.

23 On the asymmetries that defenses bring to negotiations, see Glaser and Fetter 2001: 77–9.

24 The ability to inflict the worst case improves most markedly against ever-weaker defenses as force sizes increase: the top curve in each graph leans more decidedly towards lower kill ratios as the worst case involves larger numbers of warheads. Moreover, the worst case is realized against a broader band of defenses when force sizes are small than when they are large. Note that the top curve is less skewed in Figure 4.4a than it is in 4b, more so in 4b than in 4c, and so forth.

25 The defense is not entirely disadvantaged here. Figure 4.1 indicates that the benefits of marginal increases in defensive performance grow most rapidly in the same region of the graph where survival probabilities are most sensitive to small force size increases. Whether defenses improve by the required amounts, though, is another matter.

26 Indeed, if intentions mattered to US policymakers, it was because they sought, through arms control, to reduce tension and facilitate positive relations between the opposing superpowers.

27 For critical scientific assessments given technological and physical challenges, see Kleppner *et al.* 2006.

5 ON THE OFFENSIVE AGAINST TERRORISTS: PUNISHMENT AND DENIAL STRATEGIES

1 This does not mean that groups do not compete to obtain credit for attacks. On this phenomenon in Iraq, see Filkins (2005).
2 Similarly, Hezbollah managed to "spin" into victory an absence of loss to a bigger and stronger Israeli force with its major offensive in Lebanon in the summer of 2006.
3 With Israel's assassination in 2004 of the blind, quadriplegic cleric, Sheikh Ahmad Yassin, the symbolic leader of Hamas, group leaders made oblique references to the US as a target in insisting that the US "take responsibility for this crime" (CNN Wolf Blitzer Reports 2004).
4 See Cragin and Daly (2004) for the relationship between terrorist group motives and capabilities.
5 Its decisions have been forged by "combining realistic considerations with traditional beliefs and arguments, emphasizing visionary goals but also immediate needs" (Mishal and Sela 2000: 3).
6 Hamas repeatedly used temporary ceasefires to "regroup" (Simon and Stevenson 2003).
7 Its susceptibility to punishment would increase even if Hamas saw cooperation as but a temporary concession, to end once the organization had taken control of the ministries and security forces of the Palestinian Authority. Conversely, Hamas can reduce its susceptibility to punishment if able to: (a) diversify its economic aid sources to reduce dependence on Western aid; (b) capitalize upon the fears of Western aid donors and Israel that a failure to assist a Hamas-controlled government will produce instability in the Palestinian territories and the collapse of fledgling Palestinian governing institutions; or (c) distance its political leaders from the organization's military wing (or conduct only joint operations with other organizations) to hide their own complicity in any violence.
8 The December 2005 suicide bomb attack by the Abu Musab al-Zarqawi organization on a Jordanian hotel could well have been an indirect effort to strike at Israel and Jordan because of Jordan's peace treaty with Israel. This organization has shown a greater willingness than the traditional al-Qaeda leadership to strike at Israeli targets (Whitlock 2005b).
9 In discussing his formal definition, Pape typically refers only to the objective requirement (e.g. p. 92); in applying it, he focuses on the subjective. His reliance on the subjective allows him to make questionable calls. For instance, he excludes Jordan from consideration as a US-occupied country ("it is not clear the United States would defend Jordan," p. 109), despite a long history of US support for Jordan and its government but includes, in the occupied group, the country of Uzbekistan, which retaliated recently for criticism of its rights practices by the Bush administration by ending US basing rights in the country.
10 It does appear, though, that mass-casualty terror attacks—those with primarily civilian targets that have killed 100 or more people—have been motivated more by nationalist and ethnic grievances than religious purposes over the last three decades (Asal and Blum 2005).
11 Admittedly, sometimes Pape concedes that religion is a relevant causal influence by suggesting that it is not the sole cause or can sometimes be a cause.

12 Going back in time, there were numerous instances of "occupations" involving groups with opposing religions and the occupier was a democracy in which suicide bombing did not occur: US involvement in Vietnam, for instance. Excluding these conflicts because suicide bombing was not reintroduced until the 1980s, only begs the question of why these bombings were reintroduced in those years. An answer requires deference to the role of militant Islam.

13 Consequently, US policies in Iraq make things more difficult for Israel in the occupied territories, just as Israel's policies in the occupied territories make things more difficult for the US in Iraq.

14 Notably, there is little to no popular support in the Israeli-occupied territories for arresting those responsible for terror attacks (despite declining support for armed insurrection).

15 One aspect of inter-group competition is that multiple groups have often taken responsibility for single attacks (Bloom 2005: 29). This is not to deny that Palestinian groups such as Hamas, Islamic Jihad, and the al-Aqsa Martyrs' Brigades have also cooperated by coordinating their efforts toward Israel and collaborating in joint operations.

16 The pattern appears to have been repeated in mid-2006 when, under severe pressure to soften its position toward Israel, Hamas condoned (at the very least) the kidnaping of an Israeli soldier to coerce the release of Palestinian prisoners held in Israeli custody, resulting in a major increase in Gaza hostilities.

17 True, the descriptive evidence *seems* to support a cyclical argument: Israel–Palestinian violence intensified up until Operation Defensive Shield, in which Israeli forces moved into Gaza and the West Bank in 2002; and Israel became progressively more reliant upon deadly weaponry such as helicopters and combat aircraft.

18 It can also be argued that electoral developments on the Israeli and Palestinian side do not follow patterns of reciprocity. The Hamas victory in the Palestinian legislative election of January 2006 followed a movement toward the center in Israeli politics. This feature of the conflict was expressed by President Bill Clinton's Middle East coordinator, when in commenting on the election he noted "how rarely the two sides were in sync" (Ross 2006).

19 Still, the evidence hardly validates ongoing offensive US military efforts in Iraq, if only because terrorists do not require majority support to be effective. Indeed, a quarter to a third of the public in Morocco and Indonesia still expressed trust in bin Laden; roughly a majority of people expressed support for suicide attacks on US and Western targets in Iraq in Lebanon, Jordan, and Morocco; and supportive majorities for bin Laden still existed in Jordan and Pakistan—countries that received considerable US economic and political support and are critical allies in the US anti-terror effort (Pakistan, because it is a base of Islamic radicalism and haven for Taliban and al-Qaeda exiles, and Jordan because it has a large Palestinian population and has been viewed as an anchor of stability in a turbulent and violent part of the region). In fact, over half of the Jordanian public believed that suicide attacks against civilian targets was at least sometimes justified in the defense of Islam (Pew Research Center 2005). See also Wright (2005).

20 See Jerusalem Media and Communication Centre (2005) and Palestinian Center for Policy and Survey Research (2005).

21 Of course, the poll numbers, as always, are open to interpretation. It should be said that polls also indicate that majorities opposed the arrest of Palestinians who attack Israel, blamed Israel for the impasse in negotiations, and believed that the Israeli disengagement from Gaza was in some manner a victory for the armed

resurrection. Indeed, a bare majority favored the idea of a two-state solution that would allow Israel to exist as a separate state (apart from whether Palestinians would have the right of return).

22 The fence prevented attacks along traditional corridors in the West Bank by forcing detours and subjecting attackers to detection when channeled through checkpoints (Byman 2006: 105).

23 By one estimate, between 2000 and 2005, Israel's successful attacks on Palestinian terrorists killed 114 other people (Byman 2006: 98).

24 On the advantages and disadvantages of assassination, see Richelson (2002).

25 Of course, these smaller groups could destroy themselves through internecine warfare or might simply wither away, as occurred with the Shining Path in Peru (Cragin and Daly 2004: 65).

26 Such operations appear centrally planned (given their common markings and apparent linkages), but the identities of the planners were unknown even to mid-level operatives that were involved in the attacks. The planners laid the ground-work for the attack and then moved on—perhaps to direct still other attacks. Despite incarcerating the local participants in the high-profile operations in Casablanca, Istanbul, Madrid, and (Mombasa) Kenya, investigators have apparently not learned the names of those who directed these attacks (Whitlock 2005a).

27 Given the complexity of the operations and the sophistication of the tactics, Israeli intelligence has questioned the Egyptian government's conclusion that the operations were entirely homegrown (Williams 2005). On the London attacks, see Leppard (2006).

28 For instance, see Israel Ministry of Foreign Affairs (2004).

29 Conversely, efforts to maintain a continued presence in a given location risks spreading forces thin and slow military progress.

6 A DEFENSIVE-DENIAL STRATEGY AGAINST TERRORISTS: HOMELAND SECURITY

1 Except where indicated, details on the September 11 strikes are obtained from National Commission on Terrorist Attacks Upon the United States (2004).

2 I refer to the "attacker" and "defender" for convenience. I recognize that terrorist "attackers" can be placed on the defensive and that "defending" governments can adopt offensive strategies.

3 This extra capability could give the attacker a "bonus" if it extends the amount of damage that is actually inflicted.

4 Conversely, strengthening the capability of clandestine organizations compromises their security. Illustrating this are the aggressive efforts by terrorists to advertise their exploits on the Internet, which have left trails that authorities can follow. On the "inverse" relationship between organizational "capacity" and "security" here, see McCormick and Owen (2000: 175–92). Efficiency is hurt by security because it depends upon duplication, duplicity, anonymity, autonomy, dispersal, and isolation.

5 This is true of technology too: the performance (e.g. speed, mobility, firepower) built into state-of-the-art weapons invites serious reliability and maintenance problems that can limit the weapons' ability to perform under certain conditions (e.g. weather, terrain).

6 This assumes little slack within operational capability such as an ability to replace a participant or attack team.

7 Illustrating the limits to creating invulnerable operational entities is Hamas' purported (hedging) tactic of preparing a second-team attack team to strike auto-

matically should the first become disabled (Cragin and Daly 2004: 47). This tactic reduces the risk that the operation will be compromised, by communication intercepts, when leaders order the second team to strike, yet it also increases the risk that the attack will be exposed because two teams share knowledge of the attack plan.

8   Likewise, a terrorist possessing the anthrax or smallpox virus is arguably positioned to realize a worse nightmare in Western societies and to attack with ease because a biological strain can be introduced surreptitiously into a population.

9   For a good history of the US and international terrorism, see Simon 1994.

10 This is not to deny the severe challenges to building construction, ventilation, and evacuation that could greatly increase the death toll in a terror attack (see Dwyer 2004).

11 Martyrdom is both good public relations and a useful diversion in that it shifts the focus from the suffering of the victims.

12 The single-day toll had actually been equaled (that year) in the incendiary attack on Tokyo.

13 This argument holds despite the terrorist appeal that biological weapons might have due to their usefulness as triggers for an apocalypse and the sacred connection between pestilence and "divine wrath" (on these arguments, see Koblentz 2003/4: 103).

14 In fact, terrorists could be attracted to a dirty bomb because of its economic and psychological effects *and* ability to produce fatalities from a conventional explosion.

15 These facts were obtained, respectively, through a research trip by the author to a casino and an Internet search.

16 Indeed, an exceedingly small proportion of reported terrorist incidents involved nontraditional weapons.

17 On other occasions, the Israeli government has exchanged Palestinian prisoners for captured Israeli soldiers (or their remains).

18 For their part, then, European governments have balked at the idea of shooting down a commercial passenger under any circumstances (Whitlock 2006).

19 The defender faces the problem of the Cold War-era choice between a "launch-on-attack" and a "launch-on-warning" system. With the former, the defender accepts some level of destruction in return for assuredness that innocent people will not be killed in response to a "false alarm."

20 In addition, it can remedy deficiencies in that approach, such as the discrimination inherent in passenger profiling.

21 For that matter, random defense can support an approach that aims for perfection (a perfect defense) but falls short due to inadequacies of training, planning, or technology. If random failings are "nonsystematic" errors (unpatterned lapses), they might be hard to exploit by an attacker and might occur with insufficient frequency to affect significantly an attacker's incentive to attack. (On systematic error in the field of airline security; see Szyliowicz 2004.)

22 More generally, dispersion is the principle behind the redundancy built into systems to protect against catastrophic breakdowns. With a duplicate or decentralized structure, partially disabled systems can perform with direction and coherence.

23 For this reason, multiple attacks can be used by the attacker to answer claims that it is on the decline (which has led some US observers to dismiss coordinated Iraqi insurgent assaults by charging they are signs of insurgent desperation).

24 These statistical realities are displayed in Figure A.1, which assesses the likely outcome of a set of attacks against four different levels of defense. The attacker

is assumed to pursue a hedge-targeting strategy—the attacking of some number of targets that, for reasons of assurance, exceeds the preferred (minimum) number of targets destroyed. The figure also considers the possibility, however, that an ambitious attacker aspires to complete success, that is, the destruction of all targets attacked. The figure has four essential features. First, it indicates that the attacker possesses the capability to launch some small number (1–8) of attacks (the lines in the graph). Second, the figure indicates the number of targets that can be destroyed in a set of attacks (on the horizontal axis). These numbers are cumulative so that a four indicates that *at least* four targets can be destroyed. For the sake of simplicity, the targets are assumed to have equal value. Third, the figure expresses the vulnerability of these targets in a "single-attack failure probability" (SAFP) that varies across the four graphs in the figure. Assumed equal across the target set, the SAFP can be viewed as the probability that an attack will *not* meet the attacker's standard of success (or, simply, the "strength of the defense"). Fourth, the figure indicates the "mission success probability" (on the vertical axis)—the chances that a desired number of targets will be destroyed given the SAFP and number of attacks.

25 Figure A.1d indicates, for example, that eight attacks do not promise even a risk-prone attacker (say, one willing to accept around a 0.25 mission success probability) more than one success against the best defenses (SAFP > 0.90).

26 The same cannot be said of attacks on weakly defended targets. Against weak defenses (SAFP = 0.25) in Figure A.1a, the attacker can plan, for instance, around a 0.75 probability of realizing five successes out of seven attacks (recognizing, of course, that what constitutes a "strong" or "weak" defense is relative).

27 The attacker must consider costs of the attack that could include the opportunity costs of being unable to strike again using the same tactics or to hit the same (or a related) target set under more favorable conditions.

28 This can be shown by comparing the prior results with those when Figure A.1 is "turned on its head" and calculations are conducted for four different "mission success probabilities" across a continuum of SAFP. The resulting graph-to-graph variation in the pattern for mission success probabilities of 0.25, 0.50, 0.75, and 0.90 is decidedly less than the variation in the pattern in Figure A.1a–1d for SAFPs of 0.25, 0.50, 0.75, and 0.90.

29 These increasing marginal returns for the defender are demonstrated in Figure A.2. It displays the effects on mission success probabilities of attacks of varying size, against targets with different levels of vulnerability, when the attacker desires: (a) one success, (b) two successes, (c) four successes, and (d) complete success.

30 This is demonstrated in Figure A.2d.

31 The defender must build on ever-higher levels of defense to realize the greatest marginal returns on a defensive investment as the attacker's relative tolerance for failure increases. For example, if the attacker will accept but one success out of eight attacks, the defender can realize dramatic increases in marginal returns only by augmenting a strong defense, that is, SAFP > 0.75. Still, the defender can obtain large returns by augmenting a mid-range defensive capability against moderately ambitious attackers, for example, those that strive for a two-of-four or four-of-eight success ratio (in Figure A.2b and A.2c, respectively).

32 Consequently, mission success probabilities are somewhat lower for redundant than hedge targeting given the same number of attacks, targets, and levels of defense.

33 To elaborate on this point, it is useful to return to Figure A.2d. The graph can now be understood to represent the *worst-case* outcome for the attacker (or best

case for the defender), when the failure of *any* attack will assuredly compromise *all* others. Apparent from the graph is that the probability of mission success is low for all but a small number of attacks conducted under very favorable conditions (i.e. a low SAFP).

34 By conceding some initial success to the attacker, these approaches could require stronger defenses to deter the attacker that seeks to destroy only a small number of targets.

35 For example, the consequences of an attack on US airlines are likely to be great because the public knows that the death rate in an airline attack is high and generalizes risk to the entire industry and because of the industry's national economic contribution and demand "elasticity."

36 This does not mean that ending the search for sharp objects will necessarily make the search for guns and explosives more efficient.

37 Extending the principle, bystander intervention proved extremely effective at reducing civilian casualties (even if taking the life of the bystander): significantly lower death tolls were recorded when a bomber was challenged before it could trip its explosives than when the bomber went unchallenged.

38 In this sense, a deterrence strategy highlights the value of handling the terrorist problem as a police action.

39 US House Select Committee on Homeland Security 2004: 25.

40 For pessimistic and optimistic assessments about the chances of preventing terrorists from acquiring and using nuclear weapons, see Allison (2004) and Frost (2005), respectively.

REFERENCES

Achen, C.H. and D. Snidal (1989) "Rational deterrence theory and comparative case studies," *World Politics*, 41: 143–69.

Albert, R., H. Jeong, and A. Barabasi (2000) "Error and attack tolerance of complex networks," *Nature*, 406(July 27): 378–82.

Albright, D. and C. Hinderson (2005): Unraveling the A.Q. Khan and future proliferation networks," *Washington Quarterly*, 28(2): 111–28.

Allison, G. (1971) *Essence of Decision: Explaining the Cuban Missile Crisis*, Boston, MA: Little Brown.

—— (2004) *Nuclear Terrorism: The Ultimate Preventable Catastrophe*, New York: Times Books.

Anderson, J.W. (2005) "U.S. and E.U. make plea on Iran atomic program: IAEA urged to refer case to Security Council," *Washington Post*, September 22, p. A16.

——, S. Fainaru, and J. Finer (2005) "Bigger, stronger homemade bombs now to blame for half of U.S. deaths," *Washington Post*, October 26, p. A1.

Andreas, P. (2005) "Criminalizing consequences of sanctions: embargo busting and its legacy," *International Studies Quarterly*, 49: 335–60.

Arend, A.C. (2003) "International law and the preemptive use of military force," *Washington Quarterly*, 26(2): 89–103.

Arkin, W.M. (2006) "The Pentagon preps for Iran," *Washington Post*, April 16, p. B1.

Asal, V. and A. Blum (2005) "Holy terror and mass killings? reexamining the motivations and methods of mass casualty attacks," *International Studies Review*, 7: 153–5.

Associated Press (2005a) "Iran calls E.U. nuclear proposal 'unacceptable,'" *New York Times*, August 6.

—— (2005b) "Bush gives Iran stern warning: 'all options' open if nuclear effort isn't halted, President says," *Washington Post*, August 14, p. A15.

—— (2005c) "Hamas vows to continue fight after pullout," *New York Times*, August 13.

Astorino-Courtois, A. (1998) "Clarifying decisions: assessing the impact of decision structures on foreign policy choices during the 1970 Jordanian civil war," *International Studies Quarterly*, 42: 733–54.

Atran, S. (2006) "The moral logic and growth of suicide terrorism," *Washington Quarterly*, 29(2): 127–47.

Baker, P. (2006) "Bush warns against nuclear-armed Iran: U.S. presses allies for U.N. action," *Washington Post*, January 14, p. A1.

—— and G. Kessler (2006) "U.S. campaign is aimed at Iran's leaders," *Washington Post*, March 13, p. A1.

Baldwin, D.A. (1985) *Economic Statecraft*, Princeton, NJ: Princeton University.

—— (1999/2000) "The sanctions debate and the logic of choice," *International Security*, 24: 80–107.

Ball, D. (1980) *Politics and Force Levels: The Strategic Missile Program of the Kennedy Administration*, Berkeley, CA: University of California Press.

Baram, A. (2001) "An analysis of Iraqi WMD strategy," *The Nonproliferation Review*, 8: 25–39.

Bernstein, A. (2001) "Monitoring large enrichment plants using thermal imagery from commercial satellites: a case study," *Science and Global Security*, 9: 143–63.

Blair, B.G. (1993) *The Logic of Accidental Nuclear War*, Washington, DC: Brookings Institution.

Bleiker, R. (2003) "A rogue is a rogue is a rogue: US foreign policy and the Korean nuclear crisis," *International Affairs*, 79: 719–37.

Bloom, M. (2005) *Dying to Kill: The Allure of Suicide Terror*, New York: Columbia University Press.

Blum, A., V. Asal, and J. Wilkenfeld (2005) "The psychology of WMD terrorism," *International Studies Review*, 7: 133–7.

Blustein, P. and W. Pincus (2006) "Port problems to dwarf new fears," *Washington Post*, February 24, p. A6.

Broad, W.J. and D.E. Sanger (2004) "The bomb merchant: chasing Dr. Khan's Network; as nuclear secrets emerge in Khan inquiry, more are suspected," *New York Times*, December 26, sec. 1, p. 1.

—— and —— (2005a) "Relying on computer, U.S. seeks to prove Iran's nuclear aims," *New York Times*, November 13, sec. 1, p. 1.

—— and —— (2005b) "Pakistan's nuclear black market seen as offering deepest secrets of building bomb," *New York Times*, March 21, sec. 1, p. 7.

Bush, G.W. (2002a) *The National Security Strategy of the United States of America*. Available from: http://www.whitehouse.gov/nsc/nss.pdf [Accessed April 23, 2006].

—— (2002b) *National Strategy to Combat Weapons of Mass Destruction*. Available from: http://www.whitehouse.gov/news/releases/2002/12/WMDStrategy.pdf [Accessed April 23, 2006].

—— (2006) *The National Security Strategy of the United States of America*. Available from: http://www.whitehouse.gov/nsc/nss/2006/nss2006.pdf [Accessed April 23, 2006].

Byman, D. (2003) "Al-Qaeda as an adversary: do we understand our enemy?," *World Politics*, 56: 139–63.

—— (2005) *Deadly Connections: States that Sponsor Terrorism*, Cambridge, UK: Cambridge University Press.

—— (2006) "Do targeted killings work?," *Foreign Affairs*, 85(2): 95–111.

Center for Defense Information (CDI), Federation of American Scientists, Natural Resources Council, and Union of Concerned Scientists (2001) *Toward True Security: A US Nuclear Posture for the Next Decade*, Cambridge, MA: Union of Concerned Scientists.

Cha, V. (2002) "Hawk engagement and preventive defense on the Korean Peninsula," *International Security*, 27: 40–78.

Chubin, S. and R.S. Litwak (2003) "Debating Iran's nuclear ambitions," *Washington Quarterly*, 26(4): 99–114.

CNN (2004) "Thousands mourn Hamas founder: angry Palestinians vow revenge after Israeli attack," March 22. Available from: http://edition.cnn.com/2004/WORLD/meast/03/22/yassin [Accessed April 23, 2006].

—— (Wolf Blitzer Reports) (2004) March 22. Transcript #032200CN.V67.

Coates, S. (2005) "Weak responses led to 9/11, Cheney asserts: inadequate retaliation seen in 7 cases," *Washington Post*, October 4, p. A18.

Cody, E. (2005a) "Talks deadlock over North Korea's demand for reactor. U.S. could seek sanctions at U.N.," *Washington Post*, September 16, p. A26.

—— (2005b) "N. Korea nuclear talks to adjourn without agreement," *Washington Post*, August 7, p. A14.

Coe, A.J. (2005) "North Korea's new cash crop," *Washington Quarterly*, 28(3): 73–84.

Coll, S. and S.B. Glasser (2005a) "In London, Islamic radicals found a haven," *Washington Post*, July 10, p. A1.

—— and —— (2005b) "Terrorists turn to the web as base of operations," *Washington Post*, August 7, p. A1.

—— and —— (2005c) "Zarqawi intertwines acts on ground in Iraq with propaganda campaign on the internet," *Washington Post*, August 9, p. A1.

Collier, P. and A. Hoeffler (2001) "Greed and grievance in civil war," *World Bank Policy Research Paper*. Available from: http://www.worldbank.org/research/conflict/papers/greedandgrievance.htm. [Accessed April 23, 2006].

—— and N. Sambanis (2002) "Understanding civil war: a new research agenda," *Journal of Conflict Resolution*, 46: 3–12.

Collins, S.D. (2003) "Dissuading state support of terrorism: strikes or sanctions? (an analysis of dissuasion measures employed against Libya)," *Studies in Conflict & Terrorism*, 27: 1–18.

Comras, V. (2005) "Following terrorists' money," *Washington Post*, June 4, p. A17.

Cordesman, A.H. (2003) *The Iraq War: Strategy, Tactics, and Military Lessons*, Washington, DC: Center for International and Strategic Studies.

—— (2004) *Iran's Developing Military Capabilities*, Washington, DC: Center for Strategic and International Studies.

—— (2005) *The Israeli–Palestinian War: Escalating to Nowhere*, Westport, CT: Praeger Security International.

Cortright, D. and Lopez, G.A. (2000) *The Sanctions Decade: Assessing UN Strategies in the 1990s*, Boulder, CO: Lynne Reinner.

—— and —— (2002) "Introduction: assessing smart sanctions: lessons from the 1990s," in D. Cortright and G.A. Lopez (eds) *Smart Sanctions: Targeting Economic Statecraft*, Lanham, MD: Rowman & Littlefield.

Cragin, K. and S.A. Daly (2004) *The Dynamic Terrorist Threat: An Assessment of Group Motivations and Capabilities in a Changing World*, Santa Monica, CA: RAND Corporation.

Crawford, N.C. and A. Klotz (eds) (1999) *How Sanctions Work: Lessons from South Africa*, New York: St Martin's Press.

Crenshaw, M. (1990) "The logic of terrorism: terrorist behavior as a product of strategic choice," in W. Reich (ed.) *Origins of Terrorism*, Cambridge, UK: Cambridge University.

200

Crocker, C.A., F.O. Hampson, and P. Aall (2004) *Taming Intractable Conflicts: Mediation in the Hardest Cases*, Washington, DC: United States Institute of Peace Press.

Daalder, I.H., J.M. Goldgeier, and J.M. Lindsay (2000) "Missile defence and arms control," *Survival*, 42: 6–28.

Danilovic, V. (2001) "The sources of threat credibility," *Journal of Conflict Resolution*, 45: 341–69.

Dareini, A.A. (2006) "Nuclear inspections are curbed by Iran: Russian plan still negotiable," *Washington Post*, February 6, p. A11.

Davenport, Christian (1995) "Multi-dimensional threat perception and state repression: an inquiry into why states apply negative sanctions," *American Journal of Political Science*, 39: 683–713.

David, S.R. (2003) "Fatal choices: Israel's policy of targeted killing," *Review of International Affairs*, 2: 138–58.

Davis, L.E. (1975) *Limited Nuclear Options: Deterrence and the New American Doctrine*, Adelphi Paper no. 121, London: International Institute for Strategic Studies.

Davis, P.K. and B.M. Jenkins (2002) *Deterrence and Influence in Counterterrorism*, Santa Monica, CA: RAND Corporation.

Deibert, R.J. and J.G. Stein (2002) "Hacking networks of terror," *Dialog-IO* (Spring): 1–14.

DeYoung, K. (2006) "Terrorist attacks rose sharply in 2005, State Dept. says," *Washington Post*, April 29, p. A1.

Dolnik, A. and A. Bhattacharjee (2002) "Hamas: suicide bombings, rockets, or WMD?," *Studies in Conflict and Terrorism*, 14: 109–28.

Drezner, D.W. (1999) *The Sanctions Paradox: Economic Statecraft and International Relations*, Cambridge, UK: Cambridge University.

—— (2003) "How smart are smart sanctions?," *International Studies Review*, 5: 107–10.

Duelfer, C. (2004) *Comprehensive Report of the Special Advisor to the DCI on Iraq's WMD*. Available from: https://www.cia.gov/cia/reports/iraq_wmd_2004/index.html [Accessed April 23, 2006].

Dvorak, P. (2005) "Washington Monument subtly fortified: security project completed with granite walls designed to provide terror protection," *Washington Post*, July 1, p. A1.

Dwyer, J. (2004) "Evacuation plans due for high rises in New York City," *New York Times*, 5 August, sec. 1, p. 1.

Edelstein, D.M. (2004) "Occupational hazards: why military occupations succeed or fail," *International Security*, 29: 49–91.

Eggen, D. and S. Wilson (2005) "Suicide bombs potent tools of terrorists: deadly attacks have been increasing and spreading since September 11, 2001," *Washington Post*, July 17, p. A1.

Einhorn, R.J. (2004) "A transatlantic strategy on Iran's nuclear program," *Washington Quarterly*, 27(4): 21–32.

el Deeb, S. (2005) "At least 15 killed in Hamas truck blast," *Washington Post*, September 24.

Elias, B. (2003) *Air Cargo Security*, CRS Report For Congress (Order Code RL32022). Available from: http://www.law.umaryland.edu/marshall/crsreports/crsdocuments/RL32022_09112003.pdf [Accessed April 23, 2006].

Ellis, J.D. and G.D. Kiefer (2004) *Combating Proliferation: Strategic Intelligence and Security Policy*, Baltimore, MD: Johns Hopkins University.

Erlanger, S. (2006) "Hamas leader sees no change toward Israelis," *New York Times*, January 29, sec. 1, p. 1.

Fair, C.C. and H. Haqqani (2006) "Think again: Islamist terrorism," *Foreign Policy* [online] January 30. Available from: www.foreignpolicy.com [Accessed April 23, 2006].

Fathi, N. (2005) "Iran says it will break U.N. seals placed at a nuclear plant," *New York Times*, August 2, sec. 1, p. 4.

Fearon, J.D. (1994) "Domestic political audiences and the escalation of international disputes," *American Political Science Review*, 88: 577–92.

—— (1997) "Signaling foreign policy interests: tying hands versus sinking costs," *Journal of Conflict Resolution*, 41: 68–90.

Feaver, P.D. (1997) "Neooptimists and the enduring problem of nuclear proliferation," *Security Studies*, 6: 93–125.

Filkins, D. (2005) "Profusion of rebel groups helps them survive in Iraq," *New York Times*, December 2, sec. 1, p. 1.

Finn, P. (2006a) "Iran, Russia reach tentative nuclear deal: details of venture to enrich uranium not yet set; agreement could prevent showdown at U.N.," *Washington Post*, February 27, p. A9.

—— (2006b) "Russia, Iran still talking on fuel enrichment plan: Tehran official says nuclear research will proceed," *Washington Post*, February 21, p. A10.

Flynn, S. (2004a) "The neglected home front," *Foreign Affairs*, 83(5): 20–33.

—— (2004b) *America the Vulnerable: How Our Government is Failing to Protect US from Terrorism*, New York: HarperCollins.

Frankel, G. (2005) "London subway blasts almost simultaneous, investigators conclude: timing devices, high explosives used," *Washington Post*, July 10, p. A19.

Freedman, L. (2004) *Deterrence*, Cambridge, UK: Polity Press.

—— and E. Karsh (1993) *The Gulf Conflict: 1990–1991*, London: Faber & Faber.

Frost, R.M. (2005) *Nuclear Terrorism after 9/11*, Adelphi Paper 378, Abingdon: Routledge.

Gaddis, J.L. (2004) *Surprise, Security, and the American Experience*, Cambridge, MA: Harvard University Press.

Gause, F.G. III (2005) "Can democracy stop terrorism?," *Foreign Affairs*, 84(5): 62–76.

Gellman, B., D. Linzer, and C.D. Leonnig (2006) "Surveillance net yields few suspects: NSA's hunt for terrorists scrutinizes thousands of Americans, but most are later cleared," *Washington Post*, February 5, p. A1.

Gerges, F. (2005) "Buried in Amman's rubble: Zarqawi's support," *Washington Post*, December 4, p. B3.

Glaser, C.L. and S. Fetter (2001) "National missile defense and the future of U.S. nuclear weapons policy," *International Security*, 26: 40–92.

—— and —— (2005) "Counterforce revisited: assessing the nuclear posture review's new missions," *International Security*, 30: 84–126.

Goo, S.K. (2005) "Airline security changes planned: threats reassessed to make travel easier for the public," *Washington Post*, August 13, p. A1.

Gordon, M.R. (2004) "'Catastrophic success': poor intelligence misled troops about risk of drawn-out war," *New York Times*, October 20, sec. 1, p. 1.

—— and B.E. Trainor (2006) *Cobra II: The Inside Story of the Invasion and Occupation of Iraq*, New York: Pantheon.

Gordon, P.H. (2001) "Bush, missile defense and the Atlantic alliance," *Survival*, 43(1): 17–36.

Graeme, C., S. Steven, and R. Gunaratna (2004) *Counterterrorism: A Reference Handbook*, Santa Barbara, CA: ABC CLIO.

Graham, B. (2005) "Zarqawi 'Hijacked' insurgency: U.S. general says foreign fighters now seen as main threat," *Washington Post*, September 28, p. A17.

—— and M. Allen (2001) "Bush to tell Russia U.S. will withdraw from '72 ABM Pact: announcement may come within days," *Washington Post*, December 12, p. A3.

Gray, C.S. (1986) *Nuclear Strategy and National Style*, Lanham, MD: Hamilton.

—— and K. Payne (1980) "Victory is possible," *Foreign Policy*, 39 (Summer): 14–27.

Gugliotta, G. (2005) "NASA seeks clearance to buy Russian technology: shuttle grounding prompts space station dilemma," *Washington Post*, September 16, p. A4.

Gunning, J. (2004) "Peace with Hamas? The transforming potential of political participation," *International Affairs*, 80: 233–55.

Gurr, N. and B. Cole (2000) *The New Face of Terrorism: Threats from Weapons of Mass Destruction*, London: I.B. Tauris.

Hadley, S.J. (2000) "A call to deploy," *Washington Quarterly*, 23(3): 95–108.

Harrison, M. (2006) "Bombers and bystanders in suicide attacks in Israel, 2000–2003," *Studies in Conflict and Terrorism*, 29: 187–206.

Hassner, R.E. (2006) "Fighting insurgency on sacred ground," *Washington Quarterly*, 29(2): 140–66.

Henderson, A.E. (2005) *The Coalition Provisional Authority's Experience with Economic Reconstruction in Iraq: Lessons Identified*, Special Report 138, Washington, DC: United States Institute of Peace.

Hoffman, B. (2002) "Rethinking terrorism and counterterrorism since 9/11," *Studies in Conflict and Terrorism*, 25: 303–16.

—— (2003) "The leadership secrets of Osama bin Laden," *Atlantic Monthly*, 291(3): 26–7.

—— (2006) "Insurgency and counterinsurgency in Iraq," *Studies in Conflict and Terrorism*, 29: 103–21.

—— and G.H. McCormick (2004) "Terrorism, signaling, and suicide attack," *Studies in Conflict and Terrorism*, 27: 243–81.

Hoffman, I. (2003) "Sandia sensors guard D.C., labs protect D.C. Metro," *Sandia/California News*. Available from: http://www.ca.sandia.gov/news/2003-news/030316_news.html [Accessed April 23, 2006].

Hopf, T. (1994) *Peripheral Visions: Deterrence Theory and American Foreign Policy in the Third World, 1965–1990*, Ann Arbor, MI: University of Michigan Press.

Howard, P. (2004) "Why not invade North Korea? Threats, language games, and U.S. foreign policy," *International Studies Quarterly*, 48: 805–28.

Howard, R. (2004) *Iran in Crisis? Nuclear Ambitions and the American Response*, London: Zed Books.

Hsu, S.S. (2004) "Plane that caused capitol evacuation nearly shot down," *Washington Post*, July 8, p. A1.

Hufbauer, G.C., J.J. Schott, and B. Oegg (2001) *Using Sanctions to Fight Terrorism*, Policy Brief 01–11, Institute for International Economics. Available from: http://www.iie.com/publications/pb/pb.cfm?ResearchID=79 [Accessed April 23, 2006].

Hurd, I. (2005) "The strategic use of liberal internationalism: Libya and the UN sanctions, 1992–2003," *International Organization*, 59: 495–526.

Huth, P.K. (1988) "Extended deterrence and the outbreak of war," *American Political Science Review*, 82: 423–43.

—— and B. Russett (1988) "Deterrence failure and crisis escalation," *International Studies Quarterly*, 32: 29–45.

Hymans, J.E.C. (2004) "The roots of the Washington threat consensus," in B. Glad and C.J. Nolan (eds) *Striking First*, New York: Palgrave Macmillan.

—— (2006) *The Psychology of Nuclear Proliferation: Identity, Emotions, and Foreign Policy*, Cambridge, UK: Cambridge University Press.

Israel Ministry of Foreign Affairs (2004) *Four Years of Conflict*. Available from: http://www.mfa.gov.il/mfa [Accessed April 23, 2006].

Jackson, B.A. (2001) "Technology acquisition by terrorist groups: threat assessment informed by lessons from private sector technology adoption," *Studies in Conflict and Terrorism*, 24: 183–213.

Jaeger, D.A. and M.D. Paserman (2005) "The cycle of violence? an empirical analysis of fatalities in the Palestinian–Israeli conflict," unpublished paper. Available from: http://www.ssc.wisc.edu/~andreoni/Jaeger-Paserman-01Aug2005.pdf [Accessed April 23, 2006].

Jehl, D. and E. Schmitt (2005) "Data is lacking on Iran's arms, U.S. panel says," *New York Times*, March 9, sec. 1, p. 1.

Jenkins, B.M. (1975) "International terrorism: a new mode of conflict," in D. Carlton and C. Schaerf (eds) *International terrorism and World Security*, London: Croom Helm.

—— (2002) *Countering al Qaeda: An Appreciation of the Situation and Suggestions for Strategy*, Santa Monica, CA: RAND Corporation.

Jentleson, B.W. and C.A. Whytock (2005/6) "Who 'won' Libya? The force-diplomacy debate and its implications for theory and policy," International Sucurity, 30 47–86.

Jerusalem Media and Communication Centre (2005) *Public Opinion Polls*, May 2005. Available from: http://www.jmcc.org [Accessed April 23, 2006].

Jervis, R. (1976) *Perception and Misperception in World Politics*, Princeton, NJ: Princeton University Press.

—— (1978) "Cooperation under the security dilemma," *World Politics*, 30: 167–214.

—— (1979) "Deterrence theory revisited," *World Politics*, 31: 289–324.

—— (1982/3) "Deterrence and perception," *International Security*, 7: 3–27.

—— (1984) *The Illogic of American Nuclear Strategy*, Ithaca, NY: Cornell University Press.

—— (1989) "Rational deterrence: theory and evidence," *World Politics*, 41: 183–207.

—— (2003) "The confrontation between Iraq and the US: implications for the theory and practice of detterence," *European Journal of International Relations*, 9: 315–37.

——, R.N. Lebow, and J. Stein (eds) (1985) *Psychology and Deterrence*, Baltimore, MD: Johns Hopkins University Press.

Johnston, D. and D. Jehl (2004) "C.I.A. sends terror experts to tell small towns of risk," *New York Times*, July 18, sec. 1, p. 18.

Jordan, M. (2005) "London calculates the cost of attacks: tourism and sales plummeted in July," *Washington Post*, August 16, p. A10.

Kahn, H. (1961) *On Thermonuclear War: Three Lectures and Several Suggestions*, Princeton, NJ: Princeton University Press.

Kahn, J. (2005) "China proposes a deal to end Korean nuclear standoff," *New York Times*, September 16, sec. 1, p. 4.

Kahneman, D. and A. Tversky (1979) "Prospect theory: an analysis of decision under risk," *Econometrica*, 47: 263–91.

Kam, E. (2004) "Curbing the Iranian nuclear threat: the military option," *Strategic Assessment*, 7(3). Available from: http://www.tau.ac.il/jcss/sa/v7n3p2Kam.html [Accessed April 23, 2006].

Karl, D.J. (1996/7) "Proliferation pessimism and emerging nuclear powers," *International Security*, 21: 87–119.

Kessler, G. (2005) "North Korea stands fast on nuclear energy use: Rice gets report on eve of talks," *Washington Post*, September 9, p. A21.

—— (2006a) "N. Korea sets terms for return to nuclear talks," *Washington Post*, March 9, p. A16.

—— (2006b) "Diplomats labor to renew talks with N. Korea," *Washington Post*, February 5, p. A18.

—— (2006c) "India nuclear deal may face hard sell: Rice set to defend landmark accord she orchestrated without Congress," *Washington Post*, April 3, p. A1.

—— and E. Cody (2005) "N. Korea, U.S. gave ground to make deal: long process looms on nuclear accord," *Washington Post*, September 20, p. A1.

Khong, Y.F. (1992) *Analogies at War*, Princeton, NJ: Princeton University Press.

Kier, E. and J. Mercer (1996) "Setting precedents in anarchy: military intervention and weapons of mass destruction," *International Security*, 20: 77–106.

Kirkpatrick, J. (1982) *Dictatorships and Double Standards: Rationalism and Reason in Politics*, New York: Simon & Schuster.

Kissinger, H. (2006) "The rules on preventive force: nuclear weapons, terrorism require rethinking a time-honored approach," *Washington Post*, April 9, p. B7.

Kitfield, J. (2005) "Coercion and preemption," *National Journal*, 37: 1618–25.

Kleppner, D., F.K. Lamb, and D.E. Mosher (2006) "Boost-phase defense against intercontinental ballistic missiles," *Physics Today Online*, February 5. Available from: http://www.physicstoday.org/pt/vol-57/iss-1/p30.html [Accessed April 23, 2006].

Koblentz, G.D. (2003) "Biological terrorism: understanding the threat and America's response," in A.M. Howitt and R.L. Pangi (eds) *Countering Terrorism: Dimensions of Preparedness*, Cambridge, MA: MIT Press.

—— (2003/4) "Pathogens as weapons: the international security implications of biological warfare," *International Security*, 28: 84–122.

Krauthammer, C. (2001) "Dense on missile defense," *Washington Post*, May 11, p. A45.

Krueger, A.B. and J. Maleckova (2002) "Does poverty cause terrorism?," *New Republic*, 226(24): 27–33.

Kydd, A. and B.F. Walter (2002) "Sabotaging the peace: the politics of extremist violence," *International Organization*, 56: 263–96.

LaFree, G., L. Dugan, and D. Franke (2005) "The interplay between terrorism, nonstate actors, and weapons of mass destruction: an exploration of the Pinkerton database," *International Studies Review*, 7: 148–51.

Laird, Melvin R. (2005) "Iraq: learning the lessons of Vietnam," *Foreign Affairs*, 84(6): 22–43.

Lebovic, J.H. (1990) *Deadly Dilemmas: Deterrence in US Nuclear Strategy*, New York: Columbia University Press.

—— (1996) *Foregone Conclusions: U.S. Weapons Acquisition in the Post-Cold War Transition*, Boulder, CO: Westview Press.

—— (2003) "The limits of reciprocity: tolerance thresholds in superpower conflict," *Journal of Peace Research*, 40: 139–58.

—— and W.R. Thompson (2006) "An illusionary or elusive relationship? The Arab–Israel conflict and repression in the Middle East," *Journal of Politics*, 68: 502–18.

Lebow, R.N. (1987) *Nuclear Crisis Management: A Dangerous Illusion*, Ithaca, NY: Cornell University Press.

—— (1989) "Deterrence: a political and psychological critique," in P.C. Stern, R. Axelrod, R. Jervis, and R. Radner (eds) *Perspectives on Deterrence*, New York: Oxford University Press.

—— and J.G. Stein (1989) "Rational deterrence theory: I think, therefore I deter," *World Politics*, 41: 208–24.

—— and —— (1994) *We All Lost the Cold War*, Princeton, NJ: Princeton University Press.

—— and —— (1995) "Deterrence and the Cold War," *Political Science Quarterly*, 110: 157–81.

Lepgold, J. (1998) "Hypotheses on vulnerability: are terrorists and drug traffickers coerceable?," in L. Freedman (ed.) *Strategic Coercion: Concepts and Cases*, New York: Oxford University Press.

Leppard, D. (2006) "MI5 admits: we've run out of leads on bombers," *Sunday Times* (London), January 29, p. 1.

Levi, M.A. (2003) "Off target," *New Republic*, 228(11): 13–14.

Levi, M.A. and M.E. O'Hanlon (2005) *The Future of Arms Control*, Washington, DC: Brookings Institution.

Levy, J.S. (1997) "Prospect theory, rational choice, and international relations," *International Studies Quarterly*, 41: 87–112.

Lewis, G., L. Gronlund, and D. Wright (1999) "National missile defense: an indefensible system," *Foreign Policy*, 117(Winter): 120–37.

——, T.A. Postol, and J. Pike (1999) "Why national missile defense won't work," *Scientific American*, (August): 36–41.

Lieber, K.A. and D.G. Press (2006) "The rise of U.S. nuclear primacy," *Foreign Affairs*, 85(2): 42–54.

Lindsay, J.M. and M.E. O'Hanlon (2001) *Defending America: The Case for Limited National Missile Defense*, Washington, DC: Brookings Institution.

Linzer, D. (2005a) "Iran is judged 10 years from nuclear bomb: U.S. intelligence review contrasts with administration statements," *Washington Post*, August 2, p. A1.

—— (2005b) "Iran resumes uranium work, ignores warning," *Washington Post*, August 9, p. A10.

—— (2005c) "Iran's president does what U.S. diplomacy could not: U.N. speech raises doubts about nuclear program," *Washington Post*, September 19, p. A12.

—— (2005d) "Congress faults nuclear deal with India: cooperation with New Delhi threatens U.S. nonproliferation policy, lawmakers say," *Washington Post*, September 9, p. A8.

—— (2006) "Strong leads and dead ends in nuclear case against Iran," *Washington Post*, February 8, p. A1.

—— and M. Moore (2006) "U.N. agency finds Iran noncompliant: nuclear officials to issue report today," *Washington Post*, April 28, p. A16.

—— and K. Vick (2006) "Iran delays threatened resumption of nuclear work," *Washington Post*, January 10, p. A12.

Loeb, V. (2001) "Ex-Clinton aide criticizes missile plan," *Washington Post*, July 20, p. A23.

Lopez, George A. and David Cortright (2004) "Containing Iraq: sanctions worked," *Foreign Affairs*, 83(4): 90–103.

Lynch, C. (2006a) "U.N. resolution would pressure Iran," *Washington Post*, July 21, p. A13.

—— (2006b) "Security Council pressures Iran: Iran is urged to halt uranium enrichment," *Washington Post*, March 30, p. A1.

McCormick, G.H. and G. Owen (2000) "Security and coordination in a clandestine organization," *Mathematical and Computer Modelling*, 31: 175–92.

McDermott, R. (2001) *Risk-Taking in International Politics: Prospect Theory in American Foreign Policy*, Ann Arbor, MI: Michigan University Press.

McDonald, J. (2005) "U.S. envoy: N. Korean talks 'excruciating,'" *Washington Post*, August 5.

Mahnaimi, U. (2005) "Revealed: Israel plans strike on Iranian nuclear plant," *Times On Line*, 13 March. Available from: http://www.timesonline.co.uk/article/0,,2089-1522978,00.html [Accessed April 23, 2006].

Makarenko, T. (2004) "The crime–terror continuum: tracing the interplay between transnational organised crime and terrorism," *Global Crime*, 6(1): 129–45.

Masood, S. and D. Rohde (2005) "Pakistan now says scientist did send Koreans nuclear gear," *New York Times*, August 25, sec. 1, p. 3.

Matthew, R. and G. Shambaugh (2005) "The limits of terrorism: a network perspective," *International Studies Review*, 7: 617–27.

Mearsheimer, J.J. and S.M. Walt (2003) "Iraq: an unnecessary war," *Foreign Policy*, 134 (Jan./Feb.): 51–9.

Medalia, J. (2005) *"Bunker Busters": Sources of Confusion in the Robust Nuclear Earth Penetrator Debate*, CRS Report for Congress (RL32599), Washington, DC: Congressional Research Service, The Library of Congress. Available from: http://www.fas.org/spp/starwars/crs/RL32599.pdf [Accessed April 23, 2006].

Mishal, S. and A. Sela (2000) *The Palestinian Hamas: Vision, Violence, and Co-existence*, New York: Columbia University Press.

Montgomery, A.H. (2005) "Ringing in proliferation: how to dismantle an atomic bomb network," *International Security*, 30: 153–187.

Moore, M. (2006a) "U.S. still short in Iran Security Council push: permanent members Russia, China favor less formal action on atomic program," *Washington Post*, January 22, p. A21.

—— (2006b) "Iran restarts uranium program: enhanced access for U.N. inspectors halted at facilities," *Washington Post*, February 15, p. A1.

—— and T.E. Ricks (2006) "Iranian leader warns U.S. of reprisal: Khamenei is defiant ahead of U.N. report," *Washington Post*, April 27, p. A1.

Morgan, P.M. (2003) *Deterrence Now*, Cambridge, UK: Cambridge University Press.

Moriarty, T. (2004) "Entering the valley of uncertainty: the future of preemptive attack," *World Affairs*, 167(2): 71–7.

Mueller, J. (2005) "Simplicity and spook: terrorism and the dynamics of threat exaggeration," paper presented at the Annual Meeting of the International Studies Association, Honolulu, March 2005.

Murray, W.S. (2001) "A will to measure," *Parameters*, 31(3): 134–47.

Myers, L., A. Roston, and the NBC Investigative Unit (2005) "Alleged terror financier operates in plain sight: Bush vowed to freeze his assets years ago, so why is he still in business?," June 30. Available from: http://www.msnbc. msn.com/id/8421366 [Accessed April 23, 2006].

Myre, G. (2004) "In loss of leaders, Hamas discovers a renewed strength," *New York Times*, April 25, sec. 1, p. 6.

Nakashima, E. (2005) "Thai teachers become targets in south: thousands terrorized by Muslim insurgency, government crackdown," *Washington Post*, August 12, p. A8.

National Commission on Terrorist Attacks Upon the United States (2004) *The 9/11 Commission Report*, New York: W.W. Norton & Company.

National Research Council, National Academy of Sciences (2005) *The Effects of Nuclear Earth-Penetrator and Other Weapons*. Available from: http://darwin.nap. edu/books/0309096731/html/R1.html [Accessed April 23, 2006].

Newnham, R.E. (2000) "More flies with honey: positive economic linkage in German *Ostpolitik* from Bismarck to Kohl," *International Studies Quarterly*, 44: 73–96.

—— (2004) "'Nukes for sale cheap?' Purchasing peace with North Korea," *International Studies Perspectives*, 5: 164–78.

Nitze, P. (1984/5) "Living with the Soviets," *Foreign Affairs*, 63(2): 360–74.

Nuclear Threat Initiative (NTI) (2006a) *Iran Profile*. Available from: http://www.nti. org/e_research/profiles/Iran/1819_1822.html [Accessed April 23, 2006].

—— (2006b) *North Korea Profile: Nuclear Overview*. Available from: http://www. nti.org/e_research/profiles/NK/Nuclear/index.html [Accessed April 23, 2006].

O'Hanlon, M. (2001) "Double talk on missile defense," *Washington Post*, July 31, p. A23.

O'Sullivan, M.L. (2003) *Shrewd Sanctions: Statecraft and State Sponsors of Terrorism*, Washington, DC: Brookings Institution.

Ottaway, M. (2003) "Promoting democracy after conflict: the difficult choices," *International Studies Perspectives*, 4: 314–22.

Palestinian Center for Policy and Survey Research (2005) *Poll Number 15* (March 2005). Available from: http://www.pcpsr.org/survey/polls/2005/p15e.pdf [Accessed April 23, 2006].

Pape, R.A. (1997) "Why economic sanctions do not work," *International Security*, 22: 90–136.

—— (2003) "The strategic logic of suicide terrorism," *American Political Science Review*, 97: 343–61.

—— (2005) *Dying to Win: The Strategic Logic of Suicide Terrorism*, New York: Random House.

Paris, R. (1997) 'Peacebuilding and the limits of liberal internationalism,' *International Security*, 22: 54–89.

Payne, K. (2005) "The Nuclear Posture Review: setting the record straight," *Washington Quarterly*, 28(3): 135–51.

Perito, R.M. (2005) *The Coalition Provisional Authority's Experience with Public Security in Iraq: Lessons Identified*, Special Report 137, Washington, DC: United States Institute of Peace.

Pew Research Center (2005) *Pew Global Attitudes Project*. Available from: http://pewglobal.org/reports/display.php?ReportID=248 [Accessed April 23, 2006].

Pillar, P.R. (2004) "Intelligence," in A.K. Cronin and J.M. Ludes (eds) *Attacking Terrorism: Elements of a Grand Strategy*, Washington, DC: Georgetown University Press.

—— (2006) "Intelligence, policy, and the war in Iraq," *Foreign Affairs*, 85(2): 15–27.

Pincus, W. (2005a) "Pentagon revises nuclear strike plan: strategy includes preemptive use against banned weapons," *Washington Post*, September 11, p. A1.

—— (2005b) "Pentagon may have doubts on preemptive nuclear moves," *Washington Post*, September 19, p. A5.

Pinkston, D.A. and A.F. Diamond (2005) *Special Report on the Shutdown of North Korea's 5MW(e) Nuclear Reactor*, Center for Nonproliferation Studies, Monterey Institute of International Studies. Available from: http://cns.miis.edu/pubs/week/pdf/050428.pdf [Accessed April 23, 2006].

Pipes, R. (1980) "Soviet global strategy," *Commentary*, 69(4): 31–9.

—— (1984) "How to cope with the Soviet threat: a long-term strategy for the West," *Commentary*, 78(2): 13–30.

Pollack, K. and R. Takeyh (2005) "Taking on Tehran," *Foreign Affairs*, 84(2): 20–34.

Posen, B.R. (1997) "U.S. security policy in a nuclear-armed world or: what if Iraq had nuclear weapons?," *Security Studies*, 6(3): 1–31.

Post, J.M. (2005) "The psychology of WMD terrorism," *International Studies Review*, 7: 148–51.

Powell, Robert (2003) "Nuclear deterrence theory, nuclear proliferation, and national missile defense," *International Security*, 27: 86–118.

Quester, G.H. (2005) "If the nuclear taboo gets broken," *Naval War College Review*, 58(2): 70–91.

Radu, M. (2001) "The rise and fall of the PKK," *Orbis*, 45: 47–64.

Reuters (2006) "Iran says Russian atomic proposal could still fly," *New York Times*, April 30.

Richburg, K.B. (1993) "Forces leaving Somalia see job well done," *Washington Post*, May 4, p. A1.

Richelson, J.T. (2002) "When kindness fails: assassination as a national security option," *International Journal of Intelligence and Counterintelligence*, 15: 243–74.

—— (2006) *Spying on the Bomb*, New York: W.W. Norton & Co.

Ronfeldt, D. and J. Arquilla (2001) *Networks and Netwars: The Future of Terror, Crime, and Militancy*, Santa Monica, CA: RAND Corporation.

Rosendorff, P. and T. Sandler (2004) "Too much of a good thing? The proactive response dilemma," *Journal of Conflict Resolution*, 48: 657–71.

Ross, D. (2006) "Give Hamas nothing for free," *Washington Post*, February 5, p. B7.

Sagan, S.D. (1995) *The Limits of Safety: Organizations, Accidents, and Nuclear Weapons*, Princeton, NJ: Princeton University Press.

—— (1996/7) "Why do states build nuclear weapons? Three models in search of a bomb," *International Security*, 21: 54–86.

—— (2000) "The commitment trap: why the United States should not use nuclear threats to deter biological and chemical weapons attacks," *International Security*, 24: 85–115.

—— and K.N. Waltz (1995) *The Spread of Nuclear Weapons: A Debate*, New York: W.W. Norton.

Salama, S. and K. Ruster (2004) *A Preemptive Attack on Iran's Nuclear Facilities: Possible Consequences*, Center for Nonproliferation Studies, Monterey Institute of International Studies. Available from: http://cns.miis.edu/pubs/week/040812.htm [Accessed April 23, 2006].

Sambanis, N. (2002) "A review of recent advances and future directions in the quantitative literature on civil war," *Defence and Peace Economics*, 13: 215–43.

Samore, G. (2003) "The Korean nuclear crisis," *Survival*, 45(1): 7–24.

Saunders, P.C. (2003) *Military Options for Dealing with North Korea's Nuclear Program*, Center for Nonproliferation Studies, Monterey Institute of International Studies. Available from: http://cns.miis.edu/research/korea/dprkmil.htm [Accessed April 23, 2006].

Sayigh, Y. (1998) "A non-state actor as coercer and coerced: the PLO in Lebanon, 1969–1976," in L. Freedman (ed.) *Strategic Coercion: Concepts and Cases*, New York: Oxford University Press.

Schaub, G., Jr. (1998) "Compellence: resuscitating the concept," in L. Freedman (ed.) *Strategic Coercion: Concepts and Cases*, New York: Oxford University Press.

Schelling, T.C. (1960) *The Strategy of Conflict*, London: Oxford University Press.

—— (1966) *Arms and Influence*, New Haven, CT: Yale University Press.

Schlesinger, J.R. (1975) *Briefing on Counterforce Attacks*, hearing before US Senate Subcommittee on Arms Control. International Law and Organization of the Committee on Foreign Relations. Washington, DC: GPO.

Schneider, J. and P. Schneider (2002) "The Mafia and al-Qaeda: violent and secretive organizations in comparative and historical perspective," *American Anthropologist*, 104: 776–82.

Schweitzer, Y. (2004) "Neutralizing terrorism-sponsoring states: the Libya 'model,'" *Strategic Assessment* 7(1). Available from: http://www.tau.ac.il/jcss/sa/v7n1p3 Sch.html [Accessed April 23, 2006].

—— and S. Shay (2003) *The Globalization of Terror: The Challenge of al-Qaeda and the Response of the International Community*, New Brunswick, NJ: Transaction Publishers.

Sciolino, E. and D. Van Natta Jr (2005) "With no leads, British consult allies on blasts," *New York Times*, July 11, sec. 1, p. 1.

Semple, K. (2006) "Iraqi death toll rises above 100 per day, U.N. says," *New York Times*, July 19, p. A1.

Seng, J. (1997) "Less is more: command and control advantages of minor nuclear states," *Security Studies*, 6(4): 50–92.

Shambaugh, D. (1999/2000) "China's military views the world," *International Security*, 24: 52–79.

Shikaki, Khalil (2006) *Willing to Compromise: Palestinian Public Opinion and the Peace Process*, Special Report 158, Washington, DC: United States Institute of Peace.

Shultz, R.H. and A. Vogt (2003) "It's war! Fighting post-11 September global terrorism through a doctrine of preemption," *Studies in Conflict and Terrorism*, 15: 1–30.

Simon, J.D. (1994) *The Terrorist Trap: America's Experience with Terrorism*, Bloomington, IN: Indiana University Press.

Simon, S. and D. Benjamin (2000) "America and the new terrorism," *Survival*, 42(1): 59–75.

—— and J. Stevenson (2003) "Confronting Hamas," *National Interest*, 74(Winter): 59–69.

Slevin, P. (2006) "Cash flow to Hamas is more restricted, deeper underground," *Washington Post*, February 19, p. A23.

Slocombe, W.B. (1999) *Remarks to the Center for Strategic and International Studies Statesmen's Forum*, Washington, DC, November 5. Available from: http://www.csis.org/html/sf991105Slocombe.html [Accessed August 14, 2001].

—— (2000) "The administration's approach," *Washington Quarterly*, 23(3): 79–85.

Snyder, G.H. (1961) *Deterrence and Defense: Toward a Theory of National Security*, Princeton, NJ: Princeton University Press.

Solingen, E. (1994) "The political economy of nuclear restraint," *International Security*, 19: 126–69.

—— (2001) "Middle East denuclearization? Lessons from Latin America's Southern Cone," *Review of International Studies*, 27: 375–94.

Sorli, M.E., N.P. Gleditsch, and H. Strand (2005) "Why is there so much conflict in the Middle East?," *Journal of Conflict Resolution*, 49: 141–65.

Squassoni, S.A. and A. Feickert (2004) *Disarming Libya: Weapons of Mass Destruction*, CRS Report for Congress. Available from: http://www.fas.org/spp/starwars/crs/RS21823.pdf [Accessed April 23, 2006].

Stein, J.G. (1992) "Deterrence and compellence in the Gulf, 1990–91," *International Security*, 17: 147–79.

Stephens, A. (2006) "With Israeli pullout, Gazans' support for groups using terrorism drops," *Program on International Policy Attitudes*, January 11. Available from: http://www.worldpublicopinion.org [Accessed April 23, 2006].

Szyliowicz, J.S. (2004) "Aviation security: promise or reality?," *Studies in Conflict and Terrorism*, 27: 47–63.

Takeyh, R. (2004/5) "Iran builds the bomb," *Survival*, 46(4): 51–64.

—— and N. Gvosdev (2002) "Do terrorist networks need a home?," *Washington Quarterly*, 25(3): 97–108.

Tannenwald, N. (2005) "Stigmatizing the bomb: origins of the nuclear taboo," *International Security*, 29: 5–49.

Thomas, W. (2000) "Norms and security: the case of international assassination," *International Security*, 25: 105–33.

Trager, Robert F. and Dessislava P. Zagorcheva (2005) "Deterring terrorism: it can be done," *International Security*, 30: 87–123.

UN Security Council (2004) "Recent attacks tragic reminders fight against terrorism far from over, Security Council told," *SC/8184 Press Release*, September 13. Available from: http://www.un.org/News/Press/docs/2004/sc8184.doc.htm [Accessed April 23, 2006].

US Commission on the Intelligence Capabilities of the United States Regarding Weapons of Mass Destruction (2005) *Report to the President*. Available from: http://www.wmd.gov/report/index.html [Accessed April 23, 2006].

US DOD (Department of Defense) (2002) *Nuclear Posture Review*. Available from: http://www.globalsecurity.org/wmd/library/policy/dod/npr.htm [Accessed April 20, 2006].

——, Joint Chiefs of Staff (2005) *Doctrine for Joint Nuclear Operations*. Available from: http://www.globalsecurity.org/wmd/library/policy/dod/jp3_12fc2.pdf [Accessed April 23, 2006].

——, Chairman Joint Chiefs of Staff (2006b) *National Military Strategic Plan for the War on Terrorism*. Available from: http://www.defenselink.mil/qdr/docs/2005-01-25-Strategic-Plan.pdf [Accessed April 23, 2006].

—— (2006a) *Quadrennial Defense Review Report*. Available from: http://www.defenselink.mil/qdr/report/Report20060203.pdf [Accessed April 20, 2006].

US GAO (General Accounting Office) (2006) *Aviation Security: Enhancements Made in Passenger and Checked Baggage Screening, but Challenges Remain*, GAO-06-371T. Available from: http://www.gao.gov/new.items/d06371t.pdf [Accessed April 23, 2006].

US House Select Committee on Homeland Security (2004) *Winning the War on Terror*. Available from: http://www.fas.org/irp/congress/2004_rpt/hscdem.pdf [Accessed August 15, 2006].

US National Intelligence Council (1999) *Foreign Missile Developments and the Ballistic Missile Threat to the United States through 2015*. Available from: http://www.fas.org/irp/threat/missile/nie99msl.htm [Accessed April 23, 2006].

US Senate Select Committee on Intelligence (2004) *Report on the U.S. Intelligence Community's Prewar Intelligence Assessments on Iraq*. Available from: http://intelligence.senate.gov/iraqreport2.pdf [Accessed April 23, 2006].

VandeHei, J. and B. Graham (2006) "Pakistan braces as Bush starts visit: president brushes aside security concerns to pay tribute to key ally on home turf," *Washington Post*, 4 March, p. A10.

Vick, K. (2006) "A bomb-builder, 'out of the shadows': Syrian linked to al Qaeda plots describes plan to attack cruise ship in Turkey," *Washington Post*, February 20, p. A1.

Wagner, R.H. (2000) "Bargaining and war," *American Journal of Political Science*, 88: 469–84.

Walsh, J. (2003) "Nuclear terrorism: risks, consequences, and response," in A.M. Howitt and R.L. Pangi (eds) *Countering Terrorism: Dimensions of Preparedness*, Cambridge, MA: MIT Press.

Washington Post (2005) "Report details terrorist attacks," July 6, p. A4.

Weisman, J. (2006) "Projected Iraq war costs soar: total spending is likely to more than double, analysis finds," *Washington Post*, April 27, p. A16.

White, G. and K. Khan (2006) "Attacks strain efforts on terror: alliance is tested by incidents along Afghan frontier," *Washington Post*, January 23, p. A1.

White, J.R. (2002) *Terrorism: An Introduction*, Stamford, CT: Wadsworth.

Whitlock, C. (2005a) "Terror probes find 'the hands, but not the brains': attackers often caught as masterminds flee," *Washington Post*, July 11, p. A10.

—— (2005b) "Amman bombings reflect Zarqawi's growing reach," *Washington Post*, November 13, p. A1.

—— (2006) "E.U.'s patchwork of policies leaves it vulnerable to 9/11-style attack," *Washington Post*, January 18, p. A1.

Wilkening, D.A. (2000) *Ballistic-Missile Defence and Strategic Stability*, Adelphi Paper no. 334, London: International Institute for Strategic Studies.

Williams, D. (2005) "Egypt gets tough in Sinai in wake of resort attacks: officials focus on local group: others see signs of outside aid," *Washington Post*, October 2, p. A24.

Wilson, S. (2005a) "Pullout focuses Israel on its future," *Washington Post*, August 13, p. A1.

—— (2005b) "Israel warns of attacks on Hamas leaders: defense minister says rocket fire must stop," *Washington Post*, September 28, p. A16.

Wolfowitz, P. (2001) *Prepared Testimony on Ballistic Missile Defense to the Senate Armed Services Committee*, July 12. Available from: http://www.defenselink.mil/speeches/2001/s20010712-depsecdef.html [Accessed April 23, 2006].

Woods, K., J. Lacey, and W. Murray (2006) "Saddam's delusions: the view from the inside," *Foreign Affairs*, 85(3).

Worth, R.F. and J. Glanz (2006) "Oil graft fuels the insurgency, Iraq and U.S. says," *New York Times*, February 5, sec. 1, p. 1.

Wright, R. (2005) "Support for bin Laden, violence down among Muslims, poll says," *Washington Post*, July 15, p. A13.

Zussman, A. and N. Zussman (2005) *Assassinations: Evaluating the Effectiveness of a Counterterrorism Policy Using Stock Market Data*, unpublished paper. Available from: http://www.arts.cornell.edu/econ/azussman/assassinations.pdf [Accessed April 23, 2006].

INDEX

Page numbers in *italic* type indicate figures.

Abbas, Mahmoud 118, 121
Afghanistan: Bin Laden 120, 133; fall of Taliban 106, 124; increase in terror attacks 9; issues for constitutional debate 58; and al-Qaeda's campaigns 118; Taliban's ethnic ties with 125; US backing of Islamic militants against Soviets 6; US cruise missile attacks 107–8; US war on 37, 69, 118, 124, 139, 145, 178, 179; war-lords 55, 58
Africa 120, 139; al-Qaeda's attacks on US embassies 105, 107–8, 147, 152
Ahmadinejad, Mahmoud 42, 103
aid: incentives for compliance 67; sanctions 60
Aideed, Mohamed 133
aircraft: flight security measures around Washington, DC 162; measures to reduce vulnerability 162, 172; in September 11 attacks 149, 150, 158; smuggling of bombs onto 158; US management of security 175
airlines: effects of September 11 events 161; protection against terrorism 171; screening of passengers 165, 172
airports: Palestinian terror attacks on 157; screening of luggage 158, 161; screening of traffic through 173, 174; security violations triggering defenses 164–5; visible security presence 166
Albright, D. 63–4
alert systems 165, 172, 174
Algerian resistance (1950s) 125
Allison, G. 78
Amman 130, 135
anti-ballistic missiles (ABM) 17
al-Aqsa Martyrs' Brigades 119, 126

Arafat, Yasser: complicity in terror attacks on Israel 119; influence with militants in Jordan 118, 119–20; Israel's confrontation of 121; political victory of Karameh 107; rejection of Camp David proposals 130
Argentina 120
arms control agreements 102
arms race 94
Asia: US special forces activities 139
assassinations 54; in approach to attacking terrorism 139, 175, 178; Israeli operations 129, 132–3; momentous effects in history 157; opportunities against government leaders 180–1
assured destruction (AD) 11–12, 16, 24, *25*; criticisms 18, 21, 82, 83; danger of activating option of 57; placement of US missiles for 13; to deter rogue states 25, 75, 87; US aims in retaliatory attack on Soviets 149–50; US estimates of destruction 28; views of proponents 27, 73
Atlanta: bombing during 1996 Olympics 95
atomic bomb: Hiroshima 156
Atran, S. 115
Atta, Mohamed 153
Aum Shinrikyo terror attack 155
automated response 76
"axis of evil" 1, 7
Ayyash, Yahiya 134–5, *135*

Baghdad 30, 77, 133–4, 159
Baldwin, D.A. 68–9
banks: financing terrorism 139, 141
Barak, Ehud 128, 130